THE EARTH IN SPACE

THE EARTH IN SPACE

EDITED BY

Hugh Odishaw

LONDON
ROUTLEDGE & KEGAN PAUL

First published in Great Britain in 1968
by Routledge & Kegan Paul Ltd.,
Broadway House, 68–74 Carter Lane,
London E.C.4

Printed in Great Britain by Photolithography
Unwin Brothers Limited, Woking and London

SBN 7100 6147 1

The Authors

W. Ian Axford, Professor of Physics and Applied Electrophysics, University of California at San Diego.

Sidney A. Bowhill, Professor of Electrical Engineering, University of Illinois.

Melvin Calvin, Nobel Laureate, Professor of Chemistry and Molecular Biology and Director of the Laboratory of Chemical Biodynamics, University of Berkeley.

Arthur D. Code, Professor of Astronomy and Director of the Washburn Observatory, University of Wisconsin.

Robert H. Dicke, Cyrus Fogg Brackett Professor of Physics, Princeton University.

F. D. Drake, Professor of Astronomy and Director of the Arecibo Observatory, Cornell University.

Walter M. Elsasser, Professor of Geophysics, Princeton University.

John Findlay, Deputy Director, National Radio Astronomy Observatory at Greenbank.

Herbert Friedman, Superintendent, Atmosphere and Astrophysics Division, U. S. Naval Research Laboratory.

Richard M. Goldstein, Research Scientist, Jet Propulsion Laboratory, California Institute of Technology.

Richard Goody, Abbott Lawrence Rotch Professor of Meteorology and Director of the Blue Hill Observatory, Harvard University.

Anton Hales, Professor of Geophysics and Chairman of the Division of Geosciences, Southwest Center for Advanced Studies.

Eugene Herrin, Professor of Geology and Director of the Dallas Seismological Observatory, Southern Methodist University.

Raymond Hide, Professor of Geophysics and Physics, Massachusetts Institute of Technology.

FRANCIS S. JOHNSON, Professor and Director, Earth and Planetary Sciences Laboratory, Southwest Center for Advanced Studies.

WILLIAM M. KAULA, Professor of Geophysics, Institute of Geophysics and Planetary Physics, University of California at Los Angeles.

GERARD P. KUIPER, Professor of Astronomy and Director of the Lunar and Planetary Laboratory, University of Arizona at Tuscon.

GORDON J. F. MACDONALD, Professor of Geophysics and Chairman of the Department of Planetary and Space Science, University of California at Los Angeles.

PETER MEYER, Professor of Physics, Laboratory for Astrophysics and Space Research, The Enrico Fermi Institute for Nuclear Studies, University of Chicago.

HUGH ODISHAW, Executive Secretary of the Division of Physical Sciences and the Geophysics Research Board and Executive Director, Space Science Board, National Academy of Sciences.

E. N. PARKER, Professor of Physics, Laboratory for Astrophysics and Space Research, The Enrico Fermi Institute for Nuclear Studies, University of Chicago.

P. J. E. PEEBLES, Professor of Astronomy, Princeton University.

COLIN S. PITTENDRIGH, Professor of Biology and Dean of the Graduate School, Princeton University.

MARTIN SCHWARZSCHILD, Eugene Higgins Professor of Astronomy, Princeton University.

HYRON SPINRAD, Professor of Astronomy, University of California at Berkeley.

HAROLD C. UREY, Nobel Laureate, Professor of Chemistry, University of California at San Diego.

WILLIAM S. VON ARX, Professor of Oceanography, Massachusetts Institute of Technology and Woods Hole Oceanographic Institution.

D. U. WISE, Professor of Geology, Franklin and Marshall College.

JOHN A. WOOD, JR., Geologist, Smithsonian Astrophysical Observatory, and Research Associate at Harvard University.

HAROLD ZIRIN, Professor of Astrophysics, California Institute of Technology, and Staff Member of the Mount Wilson and Palomar Observatories.

Preface

This book is concerned with man's environment, from the rocks beneath and the air around to stars and galaxies about him. But, in Bertrand Russell's words of forty years ago, this amounts to at least a third of all there is to know: "If our scientific knowledge were full and complete, we should understand ourselves and the world and our relation to the world. As it is, our understanding of all three is fragmentary. For the present, it is the third question, that of our relation to the world, that I wish to consider, because this brings us nearest to the problems of philosophy. We shall find that it will lead us back to the other two questions, as to the world and as to ourselves, but that we shall understand both these better if we have considered first how the world acts upon us and how we act upon the world."

The world of Russell, which we call the universe, is then the subject of the thirty essays that follow. The purpose of the book is to present current views and insights into the nature of the Earth, of the solar system, and of galaxies. Because man and the universe and his relations to the universe are part of the same whole, a separation of the three has little meaning: a better understanding of each is tied to all. In this sense, and not because a few of the essays take up aspects of life itself, the book does relate to man, his past and his future, for these are tied to the history and destiny of his planet, and hence to the origin and evolution of the solar system, and ultimately to the nature of galactic systems.

What is the solid earth like? What are the forces at the surface that shape our lives? What is the nature of the high atmosphere and the belts of radiation, now seen to be part of the Earth's magnetosphere? In what ways are the other planets and the Moon like and unlike the Earth? What is the nature and role of the Sun? Beyond Sun and Earth, what is the universe like? And in a universe built upon the same elements that make up tissue and bone, earth and air, fire and water, what leads to life in its simplest forms and to life like man's?

Not since Copernicus has man been confronted with so sharp a break in his notions of his surroundings and his relation to them. Copernicus moved the Earth to just another planetary orbit and set the Sun at the center of the scheme of things. Man, no longer the focal point, turned increasingly earthward as the Copernican view slowly and steadily took hold over the next two centuries or so.

Now we find ourselves turning outward. In our century astronomy has shown that our solar system is a speck at the edge of our galaxy and that our galaxy is but one of billions. Aside from new hypotheses about the universe, astronomy has provided a basis for rational speculation about other planets on which life might have developed. Darwin's theory of evolution and the rather recent syntheses of organic compounds from inert elements have led man to look beyond his planet.

In this expansion of man's horizons, space tools have served as a catalyst as well as carriers of those extensions of his senses which we call instruments. The very fact that his senses are no longer tied to the Earth, and that he himself can entertain the prospect of voyaging to the Moon and the nearer planets, has made the rest of the universe more palpable. At the same time it is worth noting that sounding rockets and space systems have significantly increased our body of scientific information. Our knowledge of the Earth itself, of the Moon and several planets, of the Sun, and of other stars has been greatly enhanced, and much that has been gained in these space ways could never have been obtained otherwise—for example, the new picture of the universe through X-rays, which can only be intercepted high above the absorbing atmosphere, or the large array of measurements of solar radiations and particles. Yet this is not a book about space research: data are data and those garnered by space tools become part of those gained from terrestrial observations and laboratory experiments, and all go into the mill of analysis as interpretations and insights are sought.

What is here, then, is a portrayal of man's environment—a sketch of the origin of the solar system, the nature and power of the Sun, the features of our Earth, from its molten depths to the tenuous gases held in the grip of its magnetic field, the structure and features of the Moon, the characteristics of the other planets, about which we know little but about which the next decade or two promises to prove revolutionary, interplanetary space, where a population of particles is dominated by the solar wind, the nature of the universe as revealed by optical astronomy, radio astronomy, cosmic rays, and X-rays, and lastly some aspects of life itself.

The chapters in the book stem from a series of lectures prepared for the Voice of America. I am indebted to a number of scientists for valuable advice in preparing the groundwork for this series—in particular Harry H. Hess of Princeton, Gordon J. F. MacDonald of the University of California at Los Angeles, and Herbert Friedman of the Naval Research Laboratory at Washington.

I also owe much to Mr. Theodore Wertime of the Voice of America *Forum* Series, to Mr. George A. Derbyshire, Dr. Pembroke J. Hart, Dr. Edward R. Dyer, Dr. Herbert Shepler, Dr. Frank Favorite, and Mr. Bruce Gregory for varied suggestions in general and assistance on the glossary in particular. Most helpful, too, was the secretarial, typing, and proofreading assistance of Misses Grace Marshall, Elinore Krell, M. Louise McCray, Marian Lee Scates, and Mrs. Mildred McGuire.

HUGH ODISHAW

Washington, D.C.
August 1967

Contents

PART III OTHER PLANETS AND OBJECTS

PART IV THE COSMIC ENVIRONMENT

PART V LIFE

I

THE SOLAR SYSTEM AND THE SUN

1

The Origin of the Solar System

GORDON J. F. MacDONALD

The origin of the solar system and the course of its history must be included among the great problems of natural philosophy, comparable in general interest to questions regarding the origin of life and the development of man. Indeed, the study of the origin of life and the development of man cannot be separated from cosmological considerations. An understanding of the history of the Sun, the Earth, and the other planets is required in order to fix both the conditions requisite for the development of primitive life and the change in conditions which stimulated the evolution of those forms of life now present.

Today, the solar system consists of the Sun, nine planets (many of which have extensive satellite systems), numerous smaller objects (the so-called asteroids), a large family of comets, and a mixture of dust and gas in space between the major members of the solar system. Study of this system differs from the study of stars. Many millions of stars are known to astronomy. Individual stars can be seen at different stages in their development, young stars in the process of being formed, as well as very old stars. We can thus reconstruct much of the history of a star just by observations made today, even though a star may live for many billions of years. Compared to the many millions of stars, we have only one planetary system to study. Stars with planetary systems like the Sun do not appear to be isolated freaks. There is evidence for believing that, in fact, planetary systems are common. For example, a few planets have been detected by telescopic means. We know that many stars form close partnerships, most commonly in pairs called binary stars. Both of these facts suggest that our planetary system is not unique. However, at present we cannot study these other planetary systems because the light they emit is far too weak to give detailed information.

The study of stars in our Galaxy suggested to many observers a general picture concerning how stars are born: A mass of gas and dust in the space between already existing stars is in motion, and various forces acting on the gas and dust are in balance. The main forces are those due to the gravitational attraction between the particles and the centrifugal forces due to rotation. At some stage, these forces may get out of balance, and condensation will proceed, with the dust and gas aggregating toward the center. This, then, is the initial step in a star's formation. Although the method of star formation is accepted by many scientists, there is no measure of agreement regarding how a star acquires a planetary system.

Past discussions about the origins of our solar system have been largely based on conventional astronomical observations of the motions of the planets. Today, however, we can add several new dimensions to this age-old discussion of planetary origins and constitutions. We owe our enlarged insight—which consists more of puzzling new questions than of firm answers—first to optical, ground-based observations (supplemented in recent years by ground-based observations of the planets in the radio-radar and infrared parts of the spectrum) and spectroscopic analysis of planetary atmospheres.

Stimulating this new ground-based assault is a second factor, our recently won space capability, some salient results of which are discussed later in this chapter. The third factor contributing to our enlarged understanding is new knowledge about the Earth and about objects from space that have landed on the Earth. Progress in the last area has resulted from rapid strides in the science of seismology, and in laboratory studies of mineral behavior under the high pressures which characterize the Earth's and other planets' deeper interiors. This progress also owes much to improvements in our theoretical ability to deal with conditions prevailing in those deep parts of the Earth and other planets that we can never hope to study directly, and to new instrumentation which allows us to determine not only the chemical composition there but also the ratio of isotopes of the elements.

While we still know very little about the earliest stages of the formation of the Sun and the planets, we do know quite accurately *when* these processes took place. The beginning of solar-system time can be determined by studying the relative abundance of elements and isotopes that are produced in radioactive decay. Radioactive elements, such as uranium, thorium, and potassium, decay or break up into lighter elements at a precisely fixed rate. By examining the number of daughter elements produced in the breakup of the parent elements, one can obtain an estimate of the time the radioactive element has been in place. The analysis is complicated in the sense that the material might contain some of the daughter elements initially.

Studies of the radioactive elements have been carried out on terrestrial rocks and also on meteorites, those objects that have come from space to be captured by the Earth. Various radioactive elements give an age for meteorites of between 4 and 4.5 billion years, while the oldest rocks found on the Earth are of the order of 3 billion years. By examining the total abundance of certain isotopes of lead, it is possible to obtain an estimate of the time that has passed since the Earth was formed, regardless of the thermal processes, mixing, and loss of material that have taken place since then. If this method is used, the Earth is found to be 4.5 billion years old. Thus, there is substantial data to indicate that the Earth was formed some 4.5 to 5 billion years ago. Although we know the time of the event rather precisely, we need much more information regarding the length of the accumulation process. Some estimates would make the time as short as a few tens of millions of years; others, a few hundred million years.

Let us scan the information provided by astronomy on the solar system and survey the conditions imposed and the possibilities revealed by these observations.

The first generalization which can be drawn from conventional astronomy is that the solar system today is highly organized. Each planet follows a monotonous journey about the Sun, which is predictable on a short time scale in every respect. With two exceptions—the innermost planet, Mercury, and the outermost, Pluto—the orbits are nearly perfect circles (i.e., ellipses of very low eccentricity), and they lie very nearly in a common plane. But was the system always so highly organized? Are the paths now traveled by the planets the same as the paths traveled 4 or 5 billion years ago when, as indicated by radioactive studies on terrestrial rocks and meteorites, planets had already assumed their present form? Is it possible to begin with the present configuration of the members of the solar system and trace changes in their orbits through time and space to the beginning? The answers are essentially that we cannot trace the history, although elaborate attempts to do this have been made; consequently, we do not know why the planetary orbits are as they are.

Any attempt to analyze these problems requires a detailed understanding of how every individual member of the solar system interacts with all others. As in many other fields of physics, we have only a primitive understanding of the long-term consequences of such a many-bodied interaction. Yet questions regarding many bodies interacting with each other through long-range forces must be answered in order to determine the solar system's history.

A second generalization which can be drawn from classical astronomical observations is that the Sun contains almost all the mass of the solar system whereas the planets possess nearly all the system's angular momentum. Rotation of the planets on their axes and the Sun on its axis contributes very little to the total angular momentum. Indeed, in terms of angular momentum per gram of matter—the angular-momentum density—the nine tiny planets, especially by virtue of their orbital motion around the Sun, possess 100,000 times more angular momentum than the massive but slowly rotating Sun. This inverse allotment of mass and angular momentum between the Sun and the planets has been a major stumbling block in all theories seeking to account for the origin and very early evolution of the solar system. The problem hinges not so much on the inequity in mass distribution (the Sun being almost a thousand times more massive than all the planets taken together) as it does on the profound difference in the distribution of angular momentum.

If the general picture of condensation of the Sun from a mass of gas and dust is correct, how can this condensation proceed so that all the mass falls to the center but leaves behind the angular momentum in the thin material which remains? The mystery is further compounded by certain irregularities in the distribution of angular momentum among various objects. The major planets, Mars, and certain classes of stars (the so-called early-type stars) all possess an angular-momentum density which is related to their mass in a simple way. On the other hand, late-type stars such as the Sun show a marked deficiency in angular momentum. Bright, massive, early-type stars rotate at a rate consistent with those observed for the large planets; late-type stars are smaller, less bright, and rotate at a much slower rate.

The consistent relationship between angular momentum and mass for many objects and the failure of the Sun to conform to this relationship raise a number of questions. It would appear that rotating masses having an angular momentum characteristic of the major planets and many stars are unstable as a single, high-density phase if their mass is less than about 3 to 4 solar masses and greater than about 0.1 to 0.01 solar mass. There would appear to be a range of masses for which the object cannot remain as a single body and maintain its angular velocity. If the object has this intermediate range of masses, it breaks up into two phases: a centrally condensed, very massive, high-density phase and one of low density but carrying all the angular momentum. Reasons why such a separation for intermediate-sized objects should take place are only vaguely understood, but the details of *how* this separation comes about certainly are not.

One solution to the angular-momentum problem is, then, to suppose that an object having a mass somewhat in excess of the present Sun begins to condense and, as it rotates, tends to flatten out into a disk. As the condensation due to the dominance of gravity over the centrifugal forces continues, a point is reached (somewhat analogous to the kind of phase instability one has as one approaches the critical point of a fluid) where matter separates into a high-density and a low-density phase. In such a picture of early solar-system evolution, the dominant forces controlling everything are the gravitational forces and the forces due to the rotation.

Another possible solution to the problem of angular momentum involves the action of magnetic forces, and it is thought by some that such magnetohydrodynamic processes could, in part, account for the distribution of mass as well as angular momentum in the solar system. In this theory, ionized gas thrown from the collapsing Sun by magnetohydrodynamic forces may have acted to slow the Sun's rotation as well as to remove gases from the system. The Sun would have contracted rapidly to about the dimensions of Mercury's present orbit. At that time, its rotation and increasing temperature produced strong magnetic fields near the surface. These fields produced a wiry rigidity in the ionized gas, which caused the rotating Sun to drag the outermost gas behind it and wind up its magnetic lines of force in disks about the Sun. The angular momentum of the Sun was then transferred to this gaseous disk along the magnetic lines of force. Along these, also, the gases from which the planets were to condense were slowly carried out. There are many difficulties and recalcitrant details to the theory. A major objection is that it provides no easy explanation for the observed dependence of the angular momentum on mass that is found among those planets not affected by other forces later in the development of the solar system, and also among many stars.

Whether momentum was transferred to the proto-planets by magnetic or hydrodynamic processes, it is clear from visual and spectroscopic observations that the planets, however formed, fall into two fundamentally different groups. One consists of smaller, dense planets near the Sun; the other, of less dense, very large planets at great distances from the Sun. Pluto is an exception to this statement, but we know very little of this object; initially, it may have been not a planet but, perhaps, a satellite of Neptune. The giant planets possess complex satellite systems, while the inner, terrestrial planets have either no satellites or one in the case of the Earth and two small ones in the case of Mars.

This separation of the giant gaseous planets from the smaller dense ones clearly illustrates that either chemical homogeneity did not exist during the earliest stages of solar-system evolution, or else it did not persist through the late history of the solar system when the planets were being born. The giant planets contain abundant quantities of gases such as ammonia, methane, hydrogen, and helium that are rare in the terrestial planets. Indeed, it may be that the composition of the giant planets such as Jupiter is very similar to the composition of the Sun. One may even speculate that Jupiter is a planet that was not quite massive enough to become a star, for the temperatures reached within its interior were insufficient to bring about the nuclear reactions that provide the energy of stars.

In addition to the mostly classical observations outlined above, orthodox discussions of the origin of the solar system usually stress the similarity between the planet-satellite system and the solar system itself. The analogy is only partly applicable, however, since in addition to the direct gravitational forces that dominate long-term evolution of the planetary system, there is another force which particularly influences the development of a satellite system. This additional gravitational interaction, which George Darwin discovered and labeled tidal friction, depends both on the fact that the planets and satellites are deformable bodies and on the existence of friction. If the planet and satellite possess no friction whatsoever, then the tides raised by the satellite on the planet and by the planet on the satellite do not influence in any way the long-term orbital motion of the satellite or the rotation of the planet. If friction accompanies the deformation—as it does in reality—then the tides will tend to slow the planet's rotational motion, provided the day is short compared with the month, and tend to move the satellite away from the planet. If, on the other hand, the day is long compared with the month, as is the case for one of the small satellites of Mars, then the tidal friction speeds up the planet and moves the satellite toward the planet. A remarkable feature of the tidal friction interaction is its very strong dependence on the distance between the satellite and the planet. When the satellite is close, the interaction is very strong. As the satellite moves away, the strength of the interaction decreases rapidly.

The Sun also raises tides on the planets, and the planets raise tides on the Sun. However, because of the large distances involved, we find the planetary orbits to be stable in that the solar tides produce very small changes over times of the order of billions of years. Even the innermost planet, Mercury, undergoes only minute changes in its path because of the tides raised within it by the

Sun. On the other hand, the tides raised by the Sun have influenced the rotation of Mercury to a remarkable degree. Analysis of the tide on the planet due to the Sun forces us to conclude that the planet's present orbital arrangement is stable with respect to this tidal interaction.

The same conclusion will not be reached regarding the solar system as a whole when we consider the effects of all the planets and the Sun on each other. While the planets are stable with respect to tidal friction interaction, certain of the satellite systems have undergone major changes. Analysis of such changes has shed valuable light on the past history of the solar system.

At present, for example, we know that the Moon is retreating from the Earth at a small but measurable rate—a retreat caused by the tidal friction interaction. Angular momentum transferred from the Earth to the Moon enables the Moon to enlarge its orbit, but the Earth, as a consequence, slows down in its rotation: our day gets longer.

If we extrapolate backward in time, we find a striking result. Suppose the present rate of tidal dissipation of the Earth's rotational energy is representative for all geological time—and it may well be if the Earth's internal structure took its present form quite early in the history of the planet. If that were indeed true, then the Moon must have been very close to the Earth just 1.5 to 2 billion years ago, a time short in contrast to the 4.5- to 5-billion-year age of the Earth. The indicated result of tidal interaction creates major problems for cosmology. Is the present Earth-Moon system itself a more recent development, not dating to the earliest stages of Earth history? Has the Moon arrived in our vicinity relatively recently, or did the Earth at one time possess a system of moons differing greatly from the present one? Indeed, analysis of tidal interaction suggests that the Moon may have been an interloper which was captured by the Earth and that the Moon traveled a very different path in the early part of the history of the solar system. Alternatively, the Moon may have been formed from several smaller objects circling the Earth, these objects having been formed in the near vicinity of the Earth at these early times. Both of these ideas are much more tenable than the old notion that the Moon was torn from the body of the Earth.

What about other satellite systems? Are they more stable than ours to tidal interaction? The complex satellite systems of Jupiter and Saturn, especially the presence of massive inner satellites, demonstrate that the frictional properties of the giant planets differ greatly from those of the Earth. If properties had been similar, then the large inner satellites of these giant planets would have moved away from the planet, perhaps sweeping up the smaller satellites that lay in their path. This has not happened, and we can thus conclude that the interiors of these planets differ dramatically from the Earth's, at least insofar as their dissipative properties are concerned. These planets must be much more nearly perfect elastic bodies than is the Earth, for they dissipate energy and transfer angular momentum to their satellites at a rate only 0.01 to 0.001 of that of the Earth. But even at these low rates of dissipation, major tidally induced changes in orbit have almost certainly taken place for certain satellites, particularly Triton, a satellite of Neptune.

The tidal forces not only enlarge a satellite's orbit, but they also reduce a planet's rotational speed (as pointed out above). While the angular momentum associated with a planet's orbital motion about the Sun is only slightly affected by tidal interaction, the angular momentum due to its rotation on its axis can undergo large decreases.

This effect explains why two of the four terrestrial planets—Mercury and Venus—have much lower rotational speeds than do the other planets and why the Earth has a low angular-momentum density. The Earth has an angular-momentum density that is less than would be expected for its mass. Indeed, if the Earth had not lost angular momentum through interaction with the Moon and the Sun, it would be rotating with an angular velocity corresponding to a period of 12 hours. Thus, the Earth would initially have had a day that was 12 hours long. The tidal effect operates differently in each case. The Earth, raising tides on the Moon, has brought the Moon into rotation so that the same face of the Moon is always turned toward the Earth. For the Moon, the length of day and the length of the month (the period required for the Moon to move about the earth) coincide. But this is not the case for Mercury, as we shall immediately see.

At the same time that Mariner IV was making its historic journey to Mars, equally historic ground-radar studies showed that Mercury spun on its axis once about every 59 days, not once every

88 days—the length of its year—as had been widely assumed. Since tidal interaction had equalized the Moon's length of day and its period of revolution about the Earth, it had been widely assumed that tidal interactions would similarly affect other members of the solar system. It was thought that tides raised on Mercury by the Sun were so great that they brought the rotation of the planet into coincidence with the revolution of the planet about the Sun. The discovery that Mercury has a period of rotation of 59 days quite convincingly showed that effects other than purely tidal interaction are at work. The orbit of Mercury, as has been noted earlier, differs from that of a perfect circle and is quite eccentric. Because of this—the Sun's tides seeking to equalize Mercury's rotation with the orbital angular velocity—we find that the orbital angular velocity varies. The tides are four times as strong when the planet is closest to the Sun, compared with only one-third as strong when the planet is furthest from the Sun. In addition to the tidal-varying torques, there are effects due to the fact that Mercury, like other planets, probably does not have a density distribution which is completely symmetrical about its axis of rotation. If this is true, there is a tendency for the largest axis of the equator to assume a preferred position with respect to the orbital motion. Indeed, the most probable rotational rate for a planet having deviations from symmetry about the axis of rotation, with a large eccentricity as is the case for Mercury, is one in which a year corresponds to 1.5 days. This is an important discovery, for it implies that no area of Mercury is permanently in the shadow; the planet keeps showing a new side to the Sun and to the Earth. What is surprising is that the optical observations have not revealed this faster rate of rotation because Mercury, unlike Venus, has quite definite surface detail.

Venus has been shown by similar radar studies to have a slight retrograde rotation; that is, its rotation is opposite to its orbital motion. The rate of retrograde rotation is very nearly that which would be expected if the planet were asymmetrical about its axis of rotation and if the tides raised by the Earth were of major importance. Thus, it would appear that the Earth, even though far distant from Venus and of relatively small mass, is effective in pulling Venus into its current period of rotation.

To this point, I have reviewed a few of the problems concerned with the origin of the solar system. The origin and history of the solar system remain as much of a problem as ever, but the new and powerful tools for exploring the solar system guarantee that in the future we will have even more questions to ask and perhaps a few answers to classical questions which have been raised. Exploration of the solar system by spacecraft has already influenced our thinking about the solar system in major ways.

Prior to the flight of Mariner IV, for example, many considered Mars to be a planet somewhat similar to the Earth. Nothing in the many thousands of visual observations and photographs made from the Earth suggested that the planet had a geologically inert surface; Earth-based observations had, in fact, suggested a contrary set of possibilities. The twenty-two closeup photographs of Mars taken by Mariner IV reveal a surface rather densely populated with impact craters up to 120 kilometers in diameter. On Mars we see a stark, moonlight visage which implies that Mars is much more like the Moon than the Earth. This was not the only discovery made by Mariner IV. Perhaps even more significant was the failure to discover the Martian equivalents of such Earth-like features as mountain chains, continents, or depressed basins. Apparently, the dominant Martian landscape has not been produced by mountain- and continent-building stresses originating within the planet, as has been the case on Earth. Thus, Mars is a very different planet from the Earth. We suspect the same to be true for Venus, but our uncertainties here are multiplied by the dense atmosphere which covers the solid surface.

The giant planets must also differ very greatly from the Earth. Thus, we can learn a great deal by the intense study of one planet, but to answer questions regarding the solar system as a whole, we must explore each of the planets and find their unique features and what these features tell us about the earliest part of the solar system's history. Only in this way can we hope to unravel the processes which led to the formation of the Sun, Earth, eight other planets, and the many other small objects that comprise our solar system.

• 2

The Sun as a Star

M. SCHWARZSCHILD

One might think that it should be a relatively easy matter to determine the major physical characteristics of the Sun. After all, the Sun is the one star really close to us, and many of its features can be observed directly with the help of astronomical instruments. However, when one looks over the whole list of directly observable phenomena, one soon finds that by far the majority of them are relevant only to the outermost layers of the Sun. It is true that these atmospheric layers contain many fascinating and spectacular phenomena of great scientific interest. Some of these phenomena even cause remarkable effects in our own Earth's atmosphere. However, the solar atmosphere contains only a minute fraction of the total mass of the Sun, and for our present topic—the physical structure of the interior of the Sun, which contains the bulk of the solar mass—the detailed understanding of the solar atmosphere is only of limited help.

Do we then have any direct observations relevant to the solar interior? Yes, but they consist only of the following four pieces of information.

First, we can measure directly from here on Earth the total brightness of the Sun. This measurement gives us the heat energy which the Sun loses every second by light radiation from its surface. This quantity is obviously of essential relevance to the solar interior because, as we shall see, it gives us a direct measure for the nuclear processes occurring in the very heart of the Sun.

Second, we know the total mass of the Sun. We know it because it is the determining factor for the orbital motions of the planets, which have been observed ever since man has started to watch the skies in a systematic way. Thus we know the total amount of matter which constitutes our Sun.

Third, we can easily determine the diameter of the Sun in angular measure and, because we know our distance from the Sun, we can derive from the angular diameter the true diameter in meters. Thus we know the total volume in which the mass of the Sun is contained.

I could state here the exact numerical values for these first three quantities. However, in our earthly measures these three numbers are so enormously large that they convey little meaning by themselves. On the other hand, we can compare these numbers for the Sun with those for other stars, at least to the degree that we are capable of deriving these data for other stars by various indirect means. Luckily, it turns out that the Sun is a quite average star. It is true that we have found stars that have masses nearly a hundred times larger than that of the Sun, and we have also found stars with masses more than a hundred times smaller than that of the Sun. Nevertheless, the major physical processes and circumstances that we find to be determining for the structure of the interior of the Sun seem to be also the determining characteristics, with few exceptions, for the majority of all other stars. It is exactly this lucky circumstance which gives such broad interest to the detailed study of the Sun as a star.

The fourth observable quantity relevant to the interior of the Sun is its chemical composition. By a careful analysis of all the atomic absorption lines which we observe in the spectrum of the Sun we have found, much to our surprise and quite in contrast to the chemical composition of the Earth, that by far the most abundant element in the Sun is hydrogen. About 70 percent of the mass of the Sun consists of hydrogen. Even more surprisingly, it has recently become clear that the bulk of the remaining 30 percent consists of helium, a rare chemical constituent on the Earth. Only 2 to 3 percent of the mass of the Sun consists of elements heavier than hydrogen and helium, and here we finally come to a similarity in the chemical composition of Sun and Earth: the relative proportions among the majority of these heavier elements turn out quite similar for the Sun and for the Earth.

At this point one might well object to my speaking of the composition of the Sun when in fact the spectroscopic observations obviously can give us only the composition of the solar atmosphere from which the light comes for our spectroscopic analysis. In principle this objection is entirely correct, and as a matter of fact we can still not be entirely certain that the Sun is homogeneous in its chemical composition throughout. Indeed, we are quite certain that nuclear processes have changed the composition in the very central core of the Sun, a point to which I shall return later on. But if we ignore for the moment this particular point, we believe we have now strong arguments that indicate that the Sun early in its life went through a phase in which strong mixing motions occurred throughout, from the center to the surface, which must have evened out any possible inhomogeneities that might have occurred in its initial formation. Accordingly, it would seem that we are not taking too great a risk if we assume that the composition of the solar interior is essentially identical with that measured on the solar surface, with the one exception caused by the nuclear processes I have already referred to.

After my stressing so strongly that we have only four pieces of information which we can determine more or less directly from observations and which have direct relevance to the interior of the Sun, it might well be wondered how astronomers dare to maintain that on the basis of such slim information we can derive the entire structure of the interior of the Sun, including such items as the temperature and density of the matter at the very center, the character and rate of the nuclear processes occurring there, the distribution of the total mass throughout its volume, and the mechanisms which carry the enormous heat energy produced in the solar core through all the intervening layers out to the surface where radiation carries it away. Such skepticism would be well justified if it were not for the one central fact that physicists are uncovering, step by step, the laws which govern all physical phenomena. The knowledge that these laws, if they are true general laws of physics, must be obeyed in every respect throughout the solar interior, in combination with the observed overall characteristics of the Sun, has enabled us gradually to come to a definite theory of the internal structure of the Sun.

It is true that the uncovering of basic physical laws is a continuing process and that presumably we never can come truly to the end of this process. However, for a specific problem such as the structure of the Sun we need to know only those physical laws which play major roles for this particular problem. For example, those laws which govern the cosmological evolution of the universe as a whole seem to be still far from clearly understood, even though the theory of General Relativity has given us a key starting point. But, luckily, these laws do not seem to be needed for the investigation of the Sun. The same seems to be true for those basic laws which make the elementary particles, such as protons and electrons, what they are; these laws are not yet clearly understood, but again are not needed for the theory of the Sun. Indeed, I think one can safely say that about twenty-five years ago the physicists succeeded in gaining a sufficiently precise understanding of those laws that govern the nuclear processes occurring in the majority of stars and that this step essentially completed the uncovering of all those principal laws of physics which govern the structure and evolution of the stars.

What then are these principal laws? The first of them is the law of the equilibrium of forces, recognized as relevant to the structure of stars nearly a hundred years ago. The applicability of the condition that the forces in the Sun must be in equilibrium is based on the following observations: ever since the first precise measurements of the diameter of the Sun have been made, not the slightest indication of changes of the diameter have been found, at least none that was larger than the possible measuring errors. The same is true for the brightness of the Sun.

Regarding the constancy of the brightness of the Sun we can go even further. Geologists have found traces of life in its most simple forms in geological layers for which the age has been determined as being about 3 billion years. It appears, therefore, that the temperature on the Earth at that time cannot have differed violently from the present average terrestrial temperature. Because the temperature on the Earth is clearly governed by the brightness of the Sun, we may conclude that the Sun has not changed in brightness by any appreciable amount during the last 3 billion years, an enormously strong indication of the stability of our Sun.

Similar evidence of stability can be obtained from other stars by precise measurements of their brightness. Such measurements have been made for a large number of stars throughout the past fifty years, and for the vast majority of them no brightness changes have been found. There are exceptions; an interesting class of stars, though only a small minority, shows regular periodic variations in brightness. Though these variations are large enough to be easily observable, they are caused, we now believe, by oscillations in which compression phases and expansion phases follow each other in a regular sequence. However, the amplitudes of these oscillations are so small in the interior of these stars that they in no way seriously affect the force-equilibrium condition on the average, even for this class of stars. There are more violent exceptions—namely, the novae and supernovae, which almost certainly represent violent explosions. For these very exceptional but also most fascinating cases a separate theory will have to be developed.

If we then concentrate on the vast majority of stable stars, such as the Sun, in what form must we apply the force-equilibrium condition? What forces do we have to consider?

It was early recognized that the temperature of the Sun even at its surface amounted to nearly 6,000 degrees on the absolute scale (Kelvin). Furthermore, it is easy to deduce from the enormous brightness of the Sun that the temperature of its deep interior must be in the millions of degrees. These high temperatures immediately tell us that the matter of the Sun is in the state of a gas, not of a liquid or a solid. This is true even though the mean density of the matter of the Sun—which we can compute easily by dividing the mass of the Sun by its volume—turns out to be just about equal to the density of water. Usually we associate liquids, not gases, with such high densities. However, at the enormous temperatures characteristic for stars, gases can be compressed to astoundingly high densities without switching into the liquid or solid state. The circumstance that we need only consider a gaseous state for the matter inside stars is an enormous simplification. Specifically, it permits us immediately to answer the question as to the main forces acting in the Sun.

It turns out that we have to consider only two major forces: the gravitational force which pulls the gas everywhere in the Sun toward its center and the pressure force which pushes the gas everywhere outward. Because the Sun has been stable over billions of years, we must conclude that these two forces exactly compensate each other. This means that the pressure in the core of the Sun must have exactly that high value that enables it to counteract the gravitational force—or the "weight"—of the outer layers of the Sun.

Indeed, we can go much further than this general statement. The force-equilibrium condition must be fulfilled not only for the Sun as a whole but for every single layer of gas, whether it be situated near the center or near the surface or anywhere in between. An imbalance of forces in any one layer would cause a motion of this layer which would fast disturb the neighboring layers successively and thus alter the structure of the entire Sun, contrary to its observed stability.

If we then consider any specific layer within the Sun, the force-equilibrium condition requires that the gas pressure at the inner surface of the layer (which pushes the layer outward) must exceed the gas pressure at the outer surface of the layer (which pushes the layer inward) by just the amount that the gravitational force pulls the whole layer inward. In this form the force-equilibrium condition gives us a very powerful tool to arrive at the mass distribution throughout the Sun. For this purpose we must, of course, combine the force-equilibrium condition with the gas law, which determines the pressure of a gas for any given temperature and density, as well as with the law of gravitation long ago discovered by Newton. By the way, the modifications of Newton's laws introduced by the theory of General Relativity are so small when applied to the stars that we may safely ignore them, except possibly in the case of the recently discovered quasars, of which we as yet understand very little.

The second basic physical law which we have to consider is that which governs the flow of heat energy from a hot region into a cold region. Clearly, this law is relevant for the interior of the Sun because its core, as we have seen, is very much hotter than its outer layers and because we see an enormous flux of energy emitted from the solar atmosphere in the form of light energy; this light emission would surely cool the atmospheric layers hopelessly rapidly if they were not steadily reheated by a heat flow from below.

For the purpose of formulating this energy-flow condition more precisely, let us again consider a definite layer in the Sun—more accurately, a spherical shell—somewhere between the center and the surface. The heat flow through the inner surface of the shell must be exactly equal to the heat flow through the outer surface of the shell since otherwise the shell would either gain or lose heat energy or, in other words, its temperature would rise or fall. Any temperature change in any layer of the Sun, however, would change the thermal structure of the Sun and would as a final consequence alter the brightness of the Sun, a consequence which is in direct contradiction to the geological evidence. For the core we have to formulate the heat-flow condition somewhat differently because there we know the nuclear processes which are the very sources of the heat flow. To keep the temperature of the core stable, the rate of heat flow emanating from the surface of the core has to be such that the loss of energy of the core equals the gain of energy of the core by the nuclear processes within it. Thus we have obtained a sharp condition for the heat-energy flow for every layer in the Sun.

To make this heat-flow condition a useful tool for the investigation of the thermal structure of the Sun, we have to consider the specific mechanisms which could produce a flow of heat energy within a star. Every one of these possible mechanisms will require a decrease of temperature step by step as we follow the flow from the core to the surface, for no automatic heat-flow mechanism exists that works in the absence of such a temperature decrease. However, exactly how much the temperature decrease has to be for every step to achieve the heat flow needed to fulfill the condition we have just discussed depends very much on the specific mechanism.

Here on Earth we are used to two common mechanisms for the transport of heat: conduction and convection. In conduction the heat is transported by the motion of individual atoms or molecules. In convection the heat is transported by hot masses of gases or liquids moving from the hot toward the cold regions while cooler masses flow in the opposite direction, resulting in a net heat transport from the hot region to the cold region. It turns out that conduction is hopelessly ineffective for the case of the Sun, owing to the gigantic distance and the gigantic amount of matter between the core and the surface. In contrast, convection turns out to be very effective in those layers which are just unstable enough to permit motions of gas masses outward and inward, though the required speed of these motions is very slow and correspondingly does not at all upset the force-equilibrium condition. For the Sun, it turns out that the layers occupying roughly the outer 20 percent of the radius are in this slightly unstable state; so through these outer layers the heat flow of the Sun is carried by the convective mechanism.

We would be in severe trouble with regard to the heat flow through the inner layers of the Sun if it were not for the existence of a third heat-transport mechanism. This mechanism is radiation, which under most circumstances here on Earth does not play a great role but is the dominant mechanism in large portions of most stars. Consider two adjacent layers in the Sun, with the inner one a little hotter than the outer one. The gases in the inner layer, owing to their higher temperature, will emit more light than those in the outer layer. Accordingly, the surface separating these two layers will be traversed by more light than that emitted in the outer layer and going inward. Thus there will result a net flux of radiation outward which is exactly equivalent to an outward energy flow. It is this radiative energy-transport mechanism which dominates the heat flow throughout the inner layers of the Sun.

Finally, we come to the laws governing those nuclear processes which produce the major energy sources within stars and which were the final laws to be uncovered by the physicists before theoretical astrophysicists could begin a logically complete—though of course not in all points certain—theory of the internal structure of the stars. Hydrogen, as we have seen, is overwhelmingly the most abundant constituent of the Sun, and, as it turns out now, it is also the major fuel for the energy source of the Sun, as for most stars. Four hydrogen atoms are just a little heavier than one helium atom. If four hydrogen atoms are fused into one helium atom, nearly 1 percent of the original mass is left over. Because mass and energy are equivalent, this mass defect appears as energy in the fusion process. It takes the form of gamma rays, which are immediately absorbed by the surrounding gas and thus transformed into ordinary heat energy. This is the fundamental process of *hydrogen burning* in the nuclear sense. Quite similarly, the nuclear *helium burning* process consists of the fusion of three helium atoms to form one just slightly lighter carbon atom.

For our application to the problem of the internal structure of the Sun it is, however, not enough to understand the exact character of the relevant nuclear processes. We have to know also at what rate these processes occur in a gas and how these rates depend on the temperature and the density of the gas. It is exactly this problem of the rates of the nuclear processes within the stars which the physicists successfully started to solve twenty-five years ago.

If we were to write down in precise mathematical equations all the physical laws that we have discussed, namely the force-equilibrium condition, the gas law for the pressure as a function of temperature and density, Newton's law of gravitation, the heat-flow condition, the heat-transport mechanisms, and finally the laws for the rates of the nuclear-energy-producing processes, we would end up with a formidable set of equations which would have to be solved for any given star for all its layers simultaneously. The difficulty of this problem in applied mathematics has seriously slowed down our progress in the development of the theory of the internal structure of the stars. By great good fortune, however, very large electronic computers have become more and more available to research astronomers during the past decade, and new numerical methods have been found that make the solution of our stellar structure problem for any given case entirely practicable and even very fast.

I think we do well to remind ourselves at this point that astronomers would not have been able in any way to carry through the fast and fascinating development of the theory of stellar structure and evolution if it had not been for the physicists who found the governing basic laws, the engineers who invented and built the modern computers, and the mathematicians who developed the modern numerical methods. Thus in recent years the theory of the stellar interior has become one of the prime examples of the positive effect of interactions between widely varying disciplines, however unorganized and unintended.

Let me now summarize the results for the Sun of this long sequence of theoretical undertakings. We have found that the total mass of the Sun is not at all evenly distributed over the volume of the Sun; on the contrary, the density in the outermost layers of the sun is extremely low (less than a millionth of that of water) and climbs steeply as we go inward, reaching a peak value at the center of about one hundred times the density of water. Similarly the temperature, starting at the surface with a value of a little below 6,000 degrees, keeps climbing steadily and rapidly to a central value of about 15 million degrees. This temperature value is much too low to produce *helium burning* at any noticeable rate. It is, however, just right to give the necessary rate of *hydrogen burning,* which thus is the sole energy source of the Sun—a statement that appears correct for probably more than two thirds of all the stars we know.

In the preceding discussion I have concentrated on the present state of the Sun. Let me add one short glance back into the past history of the Sun's life. There is various evidence that meteorites and the Earth originated about 4.5 billion years ago. Furthermore, it is hard to conceive that the Sun could have originated much before or much after the birth of the rest of the solar system. It seems reasonable then to assume that the Sun is now approximately 4.5 billion years old. The following question then arises: What changes might the Sun have undergone during this time interval? One thing seems sure: Throughout practically its entire past life, hydrogen burning must have been the major energy source of the Sun, for the cogent reason that, to the best of our present knowledge, no comparably large source of energy has been available to the Sun. This conclusion, however, is not without consequence for our theoretical investigation of the past history of the Sun.

If hydrogen burning has been going on at approximately its present rate in the core of the Sun for 4.5 billion years, the composition of the core must have substantially changed in the sense that it is now much richer in helium (which is the product of hydrogen burning) than it was originally. Indeed, detailed computations which follow the evolution of the internal structure of the Sun step by step through the past 4.5 billion years show that nearly one half of the available hydrogen fuel in the core of the Sun has now been consumed. Thus the chemical composition of the Sun has undergone a substantial change during its past life, starting with a hydrogen-rich homogeneous mixture throughout its volume and ending at its present state with a distinctly non-uniform composition, helium-rich at the center but still hydrogen-rich in all the rest. This change in the chemical composition of the Sun during its past life, according to the detailed computations, has caused noticeable changes in the structure of the Sun, but without frighteningly large consequences: from the moment when hydrogen burning first started, the diameter of the Sun has increased by about 15 percent and its brightness by about 30 percent. We may then conclude that the past history of the Sun, at least from the time when it had settled on its hydrogen-burning career, has been relatively sedate.

Let me finish with a glimpse at what may be in store for the Sun in the future. Certainly in the near future, say for 4 billion years, the sun will just continue to consume the hydrogen fuel it has left in the core. What will happen when the hydrogen is exhausted at the center? Detailed computations suggest that the answer to this question is: nothing too serious—yet. While hydrogen burning must cease in the core when its fuel is exhausted, the hydrogen burning has only to move a little further out to find ample fuel to feed on. Thus the Sun develops a new internal structure: an inactive helium core surrounded by a hydrogen-burning shell, with the main portion of the mass of the Sun (still hydrogen-rich) in turn surrounding the burning shell.

During this development, with an inactive helium core containing a steadily increasing fraction of the total solar mass, the rate of growth of the solar diameter speeds up but the increase of the brightness of the Sun remains modest. But now, after less than a billion years in this new phase of development, a rather sudden change in the evolution of the Sun seems to occur. When the inactive helium core contains approximately one third of the mass of the Sun, this core starts to contract rather rapidly while the hydrogen-rich outer portions of the Sun will expand at an accelerating pace. Worst of all, the brightness of the Sun will start to increase rapidly and in less than 100 million years will reach a value that will cause the Earth to boil, in a literal sense.

From the narrow point of view of mankind, little interest attaches to the evolutionary history of the Sun beyond this point. This narrow, man-centered point of view should, of course, not be taken by a scientist. Nevertheless, it may be wise for me to stop at this point because I would soon have to admit that all computations thus far made regarding the evolution of the Sun beyond this critical point have not yet reached the degree of mathematical completeness and physical accuracy to serve as a secure basis for further definitive predictions.

• 3

The Solar Atmosphere

HAROLD ZIRIN

As the Sun rotates about its axis every 27 days, its surface is constantly changing, within a larger, more persistent structure. The surface sloshes back and forth every 4 minutes, small granules appear and die away in 8 minutes; sunspots appear, grow, and fade out in a few weeks or months, their lifetimes punctuated by the great outbursts we call solar flares. All of this activity rises and falls in the great 11-year sunspot cycle. These are the great phenomena of the solar atmosphere, whose effects reach out to the Earth and beyond it through the solar system.

The Sun is so hot that it is completely gaseous, and therefore its surface is not hard and sharp like the Earth's. In fact, we define the surface of the Sun as that level to which we may see in integrated light—the total visible white light. It is the level in the atmosphere at which the density has dropped so low that the gas is transparent. All or most of the radiant energy may now stream outward into space. At this boundary, which we call the photosphere, a number of remarkable changes in the behavior of the solar plasma occur.

Because the density drops off sharply and the radiant energy suddenly escapes, convective currents rising from below grow into energetic shock waves. At the same time the gas in the atmosphere sloshes back and forth and up and down just like water in a bathtub. Strong magnetic fields are generated, and these combine with the motions to produce heating of the atmosphere, so that the temperature, which has dropped all the way out from the center of the Sun, rises rapidly to a million degrees.

The tenuous million-degree atmosphere, called the corona, is seen as a halo of pearly light in total eclipses, when the bright light of the surface is blocked out by the Moon. The corona reaches out as far as the Earth.

The density at the surface of the Sun falls off because the lower layers must bear the weight of the upper layers; they can only do this if the pressure is higher down below. This is called barometric equilibrium. The same phenomenon occurs in the Earth's atmosphere; the density decreases quite sharply with height. We can calculate that (at the temperature of the Sun's surface, 6,000 degrees) the density decreases to a twentieth at a height of 500 kilometers. When we look at the Sun's limb from the Earth at a distance of 150 million kilometers, it looks quite sharp to the eye as well as to the telescope. The finest telescopes can resolve only about 700 kilometers on the Sun under the best conditions.

If we look at the Sun in white light, we at once see several important features. First, the Sun is darker near the edges, so the layers we see there must be cooler. Since we cannot see so deeply into the atmosphere when we look slantwise, we conclude that the temperature is still decreasing at the height defined by the edge of the Sun. The temperature falls from 6,000 degrees at the levels which we see at the center of the Sun to about 4,500 degrees near the edge.

The second important feature we see is the granulation, a fine pattern like corn grains about 1,000 kilometers across. These grains cover the entire Sun; each grain appears, lives about 8 minutes, and breaks up or fades away. The granules appear to represent convective currents carrying heat outward from the interior. If we study the velocities of the gases in the photosphere carefully, we find that there is a larger-scale pattern, the supergranulation, which has cells around 30,000 kilometers across, in which the gases flow outward to the edges of the cell. Moreover, the gas at any point in the atmosphere rises and falls rhythmically with a period of 250 seconds and a velocity of a third of a kilometer a second (1,200 kilometers an hour).

Because of the continual outward flow in the supergranulation cells, magnetic fields accumulate at their edges. At these edges, gas pressure still is greater than the pressure of the magnetic field. But 1,000 kilometers above the granule edges, the gas pressure has decreased by 400 times, and there the magnetic fields, which do not decrease so rapidly with height, restrain and organize the motions of the ionized gases. The result is that when we look at higher levels we see a very strong cellular supergranulation structure.

How do we look at higher levels in the atmosphere? These levels are easily accessible to our line of sight, but the gases are quite transparent, so we see right through them, just as we see through our own atmosphere. In order to see the tenuous atmospheric gases, we must use a technique which permits us to look in frequencies absorbed by the gases—for example, the spectrum lines of hydrogen may be used. Another way, much older, is to take advantage of a total eclipse when we can observe the very last crescent of the Sun just as the rest of the surface is covered by the Moon. At the instant before totality a bright pink flash of light from the outer edge of the Sun is seen; that layer is therefore called the chromosphere.

When we examine the chromosphere in hydrogen or calcium light, we may see the strong supergranulation pattern. The edges of the cells form a network of higher temperature and stronger magnetic fields. If we look carefully we see rapid jets of gas, called spicules, shooting up at the edges of the cells. Their velocity is about 30 kilometers a second, and they rise about 5,000 kilometers above the surface. Although there are not many of them on the disk, when we look at the limb the foreshortening merges them into a forest. It is from these jets that material flows into the corona above, and through them flows the energy that heats the corona.

The corona is a very remarkable region. It can be studied only at eclipses or at high altitudes with coronagraphs that block out the light of the sun itself. For the corona is a million times fainter than the disk of the Sun and so is completely lost in a bright and hazy sky. We know the corona is very hot because of the spectrum lines emitted there. From the radiation we find ionized iron with 13 or more electrons removed, ionized calcium with 14 electrons missing, and so on. Such high ionization can only be produced at very high temperatures. Although the corona is transparent to ordinary light, it is opaque to radio waves longer than 5 meters, and radio observations confirm its high temperature. We can also show that it produces scintillation in the light of distant radio stars even when they are 90 degrees away in the sky, which proves that the coronal gas extends all the way to the Earth.

Because the corona is so hot, it radiates a good deal in the ultraviolet. Accordingly, when we observe the ultraviolet spectrum from rockets or satellites, the spectrum is dominated by the lines of the highly ionized coronal atoms. By observing in this region, we can get some information on the corona as it appears on the disk of the Sun, rather than just looking at the edge.

Although the corona is very hot, we often see much cooler clouds, called prominences, above the surface of the Sun. These clouds are almost transparent except in hydrogen light, but at that wavelength they are considerably brighter than the corona. They are best seen against the sky with the disk light blocked out. But they may also be seen against the disk of the Sun, where they appear dark. This is because they are darker than the disk but brighter than the sky.

When we study the positions of prominences on the Sun, we find they are located on the boundary between large magnetic regions of north and south polarity. Magnetic lines of force rise up on one side and come down on the other, and in between the field is horizontal. Since the ionized gas cannot cross the field lines, it is supported above the surface. So the prominences are accumulation of cooled-off coronal material supported against gravity by horizontal magnetic fields. If we make movies of prominences, we may see material slowly moving downward. If the prominence is near a spot group, gas flows down to the spot along arching field lines. Sometimes the magnetic field changes abruptly, and the whole prominence blows out from the Sun in a great arch.

I have so far been concerned with the quiet Sun and the behavior of the atmosphere when undisturbed by transient activity. But the most exciting occurrences on the face of the Sun are the phenomena connected with sunspots.

Sunspots are dark regions on the surface of the Sun. They occur in many sizes, from little pores a 1,000 kilometers across to giants 100,000 kilometers in diameter that may be seen with the naked eye. They occur between latitudes 5 and 40 degrees in both hemispheres, although in the last ten years there have been very few spots in the southern hemisphere. The number of spots varies cyclically, with 11 years separating successive minima. At the beginning of a cycle small spots appear at high latitudes. As time goes on, the spots grow larger and more numerous, and they also occur closer to the equator. The last spots of a cycle are quite close to the equator.

Sunspots have very strong magnetic fields; their field is ten times stronger than an alnico magnet of the best quality, and one can imagine the strength of such a magnet 100,000 kilometers across. The magnetic field is thought to suppress the convection of heat from below and thus make the sunspot cooler than its surroundings, which explains its darkness. Larger spots tend to occur in groups, with one polarity on the east side of the group and the other in the west. The polarity of spot groups in the northern and southern hemispheres is opposite. With a new cycle, the polarity of the magnetic field changes, so that it takes two cycles—or a single 22-year full cycle—to come round to the same situation again. No one can explain this remarkable cycle.

Typically, large spot groups last 2 or 3 months. Because the Sun rotates once in 27 days, we can see large spots come around several times.

What happens to the sunspot fields when the spots die? The magnetic fields are dragged out by the motions in the surface and spread over the Sun. This is helped by the fact that the Sun rotates more slowly at higher latitudes, so that fields which drift poleward lag behind and are stretched over the surface. Soon large areas of the surface are covered with weak magnetic fields of one dominant polarity. These fields are marked by long streamers in the corona, and they are even detected in interplanetary fields near the Earth.

Every once in a while—sometimes every few hours in particularly active groups—a great outburst of energy occurs in the neighborhood of a sunspot group. This is a solar flare, a truly remarkable phenomenon. Regions tens of thousands of kilometers across will brighten simultaneously in a matter of seconds. Great clouds of matter are thrown out with velocities of 500 to 1,000 kilometers a second. Flares are transparent in ordinary light. Yet if we look in the extreme ultraviolet (the most energetic part of the spectrum), a flare covering a thousandth of the surface emits more light than all the rest of the Sun. Flares are most conveniently seen in the wavelengths of hydrogen light. By limiting ourselves to those wavelengths we reject most of the light of the surface but retain most of the flare emission, making it easily visible.

At the moment of most rapid brightening, energetic pulses of X-rays are emitted that change the Earth's ionosphere so that radio signals fade out, and swarms of energetic cosmic rays are emitted that fill interplanetary space. To be sure, the biggest flares that severely disrupt the ionosphere and produce really hazardous cosmic radiation are infrequent—a few a year and only in the biggest spot groups. But even modest sunspot groups will have numerous small flares, each of which produces its own pulses of energy.

Careful observation of flares, particularly by cinematography, shows that they frequently occur in regions having a steep magnetic field gradient, and that they are most common in very complex sunspot groups with intertwined regions of different polarity. To explain how flares occur, we must explain how their energy is stored up and then released very rapidly. The underlying sunspot and granulation structure is unchanged by the flare. Although flares have a lot of energy, it is minuscule compared to the enormous thermal energy under the surface of the sun. What makes the flares important is that a great deal of their energy is organized and concentrated in the most energetic part of the spectrum.

If we study the corona above an active sunspot group, we find a relatively dense cloud of hot gas at more than 3 million degrees. Each flare or eruption throws more material upward at high velocities, and these velocities are dissipated in a general heating of the atmosphere. The sunspot magnetic fields extend high above the surface, and often we see graceful loop prominences, which occur as the hot material thrown up by the flare cools, condenses, and rains down along the curving magnetic lines of force. The hot gas in these coronal condensations emits a considerable quantity of soft X-rays; in addition, we often find hard X-rays coming from this region. Such radiation is particularly noticeable when a flare occurs just over the edge of the Sun, so we see the eruption in the atmosphere even though we don't see the flare itself. The fast-moving electrons produced in the flare are trapped in the atmospheric magnetic fields and radiate their energy in the form of X-rays.

Why do sunspots occur? This question has always fascinated astronomers. Early theories simply considered them as storms on the Sun. If we look at atmospheric structure around sunspots in hydrogen light, we see strongly curved configurations, like the curved clouds around a hurricane. We now know that these clouds are elongated because matter is forced to flow along the magnetic lines of force. And we know that the strong magnetic fields in spots suppress motion, so that the spots are rather quiet although the atmosphere above them is very turbulent.

Many theories of the sunspot cycle connect it with the Sun's differential rotation—the remarkable fact that the Sun rotates faster at the equator than it does at the poles. Some astronomers have conjectured that this unequal rotation winds up the magnetic lines of force, greatly intensifying them, until sunspots break out.

Other astronomers feel that the differential rotation is due to the spots themselves. They suppose that the inside of the Sun rotates somewhat more rapidly than the surface, which is slowed by the interaction of atmospheric magnetic fields with the interplanetary medium. The sunspots sink roots from the slowly rotating atmosphere into the interior and speed things up.

But we still don't know how the spots are produced, and we cannot see why they should return so regularly every 11 years.

We passed through a minimum of solar activity in 1964, when a new cycle was about to begin. Astronomers tried to prepare for the new cycle with a variety of new instruments to observe the phenomena. We were especially interested in rapid time sequence observations so that we could observe the evolution of fast-changing phenomena, and high-resolution observations so that we could see exactly what is going on. One important source of information is data from rockets and satellites in regions of the spectrum that do not penetrate our atmosphere, particularly the ultraviolet. In this region we may observe directly the parts of the atmosphere, such as the corona, that are transparent in the visual spectrum. Also, the more energetic ultraviolet light, particularly X-rays, reflect most closely the energetic processes in flares. So we hope, with the further development of satellite and rocket astronomy, that we shall gain new knowledge from a different point of view.

Another way in which we are gaining new knowledge about the Sun is by the study of similar activity in other stars. Although the stars are so distant that we cannot see their surfaces (they appear as points), by studying the behavior of certain lines in their spectrum we can determine if they have chromospheres or solar activity. These lines are, of course, the same strong spectrum lines in which we study the solar chromosphere and flares. We can see how often and how strongly these phenomena occur in stars of different ages and sizes, and thus place these phenomena in the proper perspective in the lifetime of a star. On the other side, by studying the phenomena in the Sun from the stellar point of view, we may explain to the stellar astronomers the meaning of these barely detectable phenomena, which we only can interpret by looking at the surface of our Sun, the only star that we really can see in two-dimensional detail.

4

The Solar Wind in Space

FRANCIS S. JOHNSON

Until about twenty years ago it was generally thought that interplanetary space was an ideal vacuum. It was recognized that, at times, the Sun ejected clouds of gas that reached the Earth and caused magnetic disturbances and auroras. However, these events were believed to occur only occasionally, and it was thought that interplanetary space was devoid of any atmosphere between events. This point of view underwent relatively rapid change following World War II, and the concepts of what is happening in interplanetary space, from the point of view of gas dynamics, are still undergoing rapid change.

Radio observations of atomic hydrogen first provided a new source of data that affected the earlier viewpoint significantly. A hydrogen atom consists of a proton and an electron, each of which is spinning, and the spin axes may be parallel or antiparallel. When the spin axes are parallel, the configuration has a higher energy than when they are antiparallel, and there is a tendency, albeit very weak, for the spin of one to reverse so that energy can be released; when this happens, the atom emits a photon of radio energy with a wavelength of 21 centimeters. Van de Hulst predicted that such emissions from hydrogen in galactic space should be observable. H. I. Ewen and E. M. Purcell, and others also, observed the effect by utilizing radio telescopes tuned to the appropriate frequency. In addition, a Doppler shift was observed for hydrogen clouds that are moving toward or away from the observer. In this way, it was recognized that there is about one hydrogen atom per cubic centimeter in our Galaxy, and the rough distribution of hydrogen in the galactic arms was deduced. For the present discussion, the important point is that galactic space is not an ideal vacuum, but instead that it contains a tenuous neutral gas that is very cold, the temperature being of the order of 100 degrees Kelvin.

The next important development concerned ionized hydrogen within the solar system. Three sources of evidence arose at about the same time and seemed to corroborate one another. Comet tails were observed always to be deflected away from the Sun by an amount that could not be explained in terms of light pressure; Von L. Biermann interpreted this observation as indicating a continual outstreaming of ionized hydrogen from the Sun, with a concentration near the Earth's orbit of about 1,000 protons per cubic centimeter and moving with a velocity of the order of 1,000 kilometers per second. The zodiacal light, a faint band of luminosity across the night sky attributable mainly to interplanetary dust (i.e., micrometeoroids), is partly polarized; this polarization was at first attributed to free electrons in interplanetary space, the required concentration being of the order of 1,000 electrons per cubic centimeter. Finally, radio whistlers, or whistling atmospherics that are observed on the Earth, were interpreted by Owen Storey as being caused by lightning strokes, the signals traveling along the magnetic field lines out into space and back to the opposite hemisphere; these required the presence in space of about 1,000 electrons per cubic centimeter. In the late 1950's, it was therefore commonly believed that interplanetary space contained the relatively high concentration of ionized hydrogen of about 1,000 protons and electrons per cubic centimeter. The main difference in views current at that time was whether the medium was relatively static, as described by Sydney Chapman, or streaming continuously outward from the Sun in the form of a solar wind, as described by Eugene Parker.

The term "solar wind" was coined by Eugene Parker to indicate the expanding outer atmosphere of the Sun. This outer atmosphere or corona undergoes what amounts to a continuous controlled explosion or expansion that causes a continual outstreaming of solar gases into interplanetary space. The phenomenon is comparable to the flow of gas out of a rocket motor. Such a flow from a rocket motor is supersonic, and in a sense the solar wind can be considered a supersonic, or even hypersonic, flow out of a set of rocket nozzles that completely covers the sun.

The analogy leads us to a pair of questions. What are the characteristic features of a rocket motor that enable it to produce supersonic flow? And what are the corresponding features of the solar atmosphere that correspond to the characteristic features of a rocket motor? First of all, in a rocket motor, heat is applied to a working fluid to raise its pressure in a combustion chamber. Only by getting the pressure high enough, relative to the pressure of the surrounding atmosphere, can the flow out of the combustion chamber reach the velocity of sound, or become sonic. This is true whether or not the exit orifice is shaped like a nozzle, but it can be accomplished only if the exit orifice is small enough to limit flow and allow a pressure buildup in the combustion chamber sufficient to accelerate the flow of sonic speed at the orifice. If rocket fuel is burned in a chamber without a constricted orifice—for example, in a gun barrel—subsonic flow will always result. And supersonic flow cannot be produced, of course, unless sonic flow is produced first.

Thus the first characteristic feature of the rocket motor is a sufficient rate of heat input into a working fluid in a confined chamber with a limited exit orifice to bring the velocity up to sonic in the orifice.

The second characteristic feature of the rocket motor is the flared nozzle in which the exhaust gases expand and cool as they accelerate out the nozzle. The rocket nozzle is a de Laval nozzle, and it bears a close physical resemblance to a Venturi nozzle, for which the flow conditions are entirely different. The gas leaving the throat of a Venturi nozzle undergoes compression instead of the expansion experienced by gas leaving the throat of a de Laval or rocket nozzle. What is the difference? Alexander Dessler has shown that the difference can be expressed in very simple terms. In the Venturi nozzle, the pressure difference simply is not sufficient to produce flow at sonic velocity at the throat and so, naturally, a supersonic expansion cannot take place, as the flow velocity never gets as high as the velocity of sound. Simply increasing the pressure at the source sufficiently to produce sonic velocity in the throat will convert the Venturi nozzle to a de Laval nozzle.

Now, how does the Sun fit into this picture? Since about the end of World War II, it has been known from radio measurements that the solar corona is hot: the gases surrounding the Sun are characterized by a temperature in excess of a million degrees. This high temperature will certainly lead to substantial escape of gas into space, and the only question is the hydrodynamic nature of the escape flow. Further, for flow radially away from the Sun, the paths of adjoining particles diverge, as they also do in a rocket nozzle. Thus two points of similarity exist between the solar corona and a rocket motor—first, the heat input to a working fluid and, second, the diverging flow in which a supersonic expansion may occur. The remaining requirement is a region of constricted flow—the equivalent of the throat of the rocket nozzle—where sonic velocity must be produced if the outflow at greater distances is to be supersonic. If such a constriction in the flow exists on the Sun, the solar corona should expand into space at supersonic velocity.

It turns out that the solar feature that corresponds to the throat of a rocket nozzzle is gravity. It is gravity that permits the pressure buildup in the lower corona that causes the acceleration of coronal gas to sonic velocity at a distance of about two or three solar radii from the center of the Sun. Beyond that, the solar corona expands and accelerates to hypersonic velocities—that is, the flow velocity becomes much greater than the average random thermal velocity of the gas particles.

Strangely, if the solar corona were hotter than about a million degrees Kelvin, gravity would not be adequate to restrict the flow enough to permit sonic velocity to be attained, and the solar wind would slow down to subsonic velocity. This condition has been described as one that would give rise to a solar breeze, but apparently it does not actually occur.

In view of its high temperature, the solar corona is virtually totally ionized. Because the Sun consists mainly of hydrogen, the solar wind consists mainly of ionized hydrogen. Observations in spacecraft show that the concentration of hydrogen ions and electrons near the Earth's orbit is about 5 to 10 ion-electron pairs per cubic centimeter, moving with a velocity of about 500 kilometers per second away from the Sun. This, then, is the solar wind. As it moves outward through the solar system, its velocity should not change much, but its concentration of particles will fall off as the inverse square of the distance to the Sun.

Because the solar wind consists of electrons and ions, it constitutes a conducting plasma. When the plasma approaches the Earth's magnetic field, currents slow in it in such a way that the magnetic field is effectively kept out of the plasma. This causes the solar wind to be deflected, and it also causes the geomagnetic field to be pushed back by the solar wind. On the side of the Earth facing the Sun, the geomagnetic field is compressed and confined. On this account, the geomagnetic field normally terminates in the direction toward the Sun at a distance of about ten Earth radii, or about 64,000 kilometers, from the Earth's center.

On the nighttime side of the Earth, the effect is different. The solar wind has already been deflected by the geomagnetic field, so it does not tend to push the magnetic field in on the nighttime side as it does on the daytime side, and the Earth thus develops a magnetic tail that streams out into space in the direction away from the Sun. The length of the tail is accentuated by another effect: Plasma from the solar wind penetrates the geomagnetic tail and effectively splits the tail lengthwise in two. The magnetic field lines therefore stream directly down the tail into space. The magnetic field lines from the south polar region are directed outward from the Earth; in the geomagnetic tail, they are therefore directed away from the Earth, and they fill the lower, or southern, half of the geomagnetic tail. The magnetic field lines from the north polar region are directed toward the Earth, and they fill the upper, or northern, half of the geomagnetic tail. There is a boundary sheet between the northern and southern halves of the geomagnetic tail, or the upper and lower halves of the tail if we think of the northern direction as upward. Above and below this boundary sheet, the magnetic field lines are oppositely directed. The boundary sheet is known as the neutral sheet, and it consists of solar plasma that has penetrated the magnetic field and split it, without the field penetrating the plasma that constitutes the neutral sheet. This splitting of the magnetic field by the plasma is a plasma effect that was not anticipated. It was recognized only after spacecraft observation made by Norman Ness showed its presence.

It is not known just how long this geomagnetic tail is. It stretches far out into space and becomes a small target for a space probe to find. It may well extend several astronomical units[1] beyond the Earth.

The solar wind is supersonic in the sense that its ordered velocity away from the Sun is greater than the average random or thermal velocity of the particles making up the solar wind. The only disturbances of a sonic type that can propagate in the plasma are hydromagnetic waves whose propagation velocity is less than the solar wind velocity, so the flow is also supersonic in this sense. Consequently, a shock wave develops in the solar wind ahead of its point of impact on the geomagnetic field. The existence of this shock wave was debated before it was first observed by spacecraft. A shock wave can exist in a neutral gas only through the effect of collisions of gas particles among themselves. If the gas density is reduced to the point where the mean free path is greater than the dimensions of the obstacle to the flow, the shock wave disappears because collisions have become too infrequent to transmit a pressure disturbance. The solar wind is so rarefied that the mean free path should be very long—greater than the distance across the earth's magnetic field. On this account, it seemed improbable that a shock wave could exist. It exists only because the particles of the solar wind interact with one another by means of electromagnetic forces, and the interaction distance is rather short. The exact nature of the interaction is not understood, but spacecraft observations make it very clear that the interaction forces do actually exist. Such an interaction can be described as one of the properties of plasmas. This and many other properties of plasmas are observed easily with spacecraft instrumentation, but only with great difficulty in the laboratory.

The shock front in the solar wind looks very much like the bow shock around a blunt or hemispherical body moving supersonically through a neutral gas. The separation distance between the geo-

[1]One astronomical unit is the mean distance between Sun and Earth—about 150 million kilometers, or 93 million miles.

magnetic field and the shock front is about four Earth radii, or 25,000 kilometers, at the center of the shock, and the separation increases as one moves away from the center, which, of course, lies on the Earth-Sun line.

It has been thought at times in the past that the flow conditions of the solar wind around the geomagnetic field would be unstable. This seemed to be an attractive hypothesis for the source of geomagnetic rapid variations, those small changes in the Earth's magnetic field that take place almost continuously with periods ranging from seconds to hours. However, both theory and observation now indicate that the flow conditions are stable and that such instabilities do not contribute to the geomagnetic rapid variations. It seems more likely that there are variations in the solar wind itself and that these cause the geomagnetic variations.

Although the solar wind pushes the Earth's magnetic field aside, it acts quite differently on the solar magnetic field. This happens because magnetic fields from sunspots or other magnetic regions on the Sun's surface extend out into that part of the solar atmosphere where the solar wind originates. As the solar wind moves out and away from the Sun, it pulls the solar magnetic field with it. It does this because it is a highly conducting medium, and if the magnetic field lines tend to slip through the plasma, currents are induced in the plasma. These currents have magnetic fields associated with them which, when added to the solar field, produce a field which is just that which would exist if the solar field were pulled out by the plasma, being unable to slip through the plasma. The net effect of the solar wind is therefore to pull any solar fields out radially.

At the same time that the solar wind pulls the solar magnetic field out radially, the Sun rotates, and this twists the otherwise radially extended field lines into a gentle spiral. Consider the solar wind that leaves the Sun from a point on the solar surface facing the Earth. In about 3 days' time, the plasma will reach the Earth. Magnetic field lines which extended through the solar surface, at the time that the particular mass of plasma that we are considering started its journey, will be stretched out to the vicinity of the Earth. However, during the 3 days that it takes to stretch the field line out from the Sun to the Earth, the Sun turns about one ninth of a revolution, or 40 degrees. Since the field lines are firmly anchored in the Sun by the high-conductivity solar gases, the field lines that have been pulled out to the Earth spiral around and enter the Sun at a solar longitude about 40 degrees to the west of the center of the Sun. The field line at the Earth makes an angle of about 50 degrees with the Sun-Earth line, being directed to a point in space 50 degrees to the west of the Sun.

I want to emphasize that the plasma moves radially outward and that it is the combined effect of the radial outward movement of the plasma and the rotation of the Sun that causes the magnetic field lines to become spiral in shape. There is a familiar analogue: if one allows water to squirt out of a garden hose and then swings the hose around to point in a different direction, the stream of water will present a curved appearance even though all parts of the stream at all times are moving radially away from the source. Because of this analogy, the angle between the interplanetary magnetic field and the direction to the Sun is frequently referred to as the "garden-hose angle."

A surprising feature of the interplanetary magnetic field is its irregularity, its small-scale structure. The large-scale features are as just described, but in addition to this there are many small-scale irregularities. Although these average to zero, the magnetic energy associated with them is a substantial fraction of the total magnetic energy. The significance of these variations is not at present appreciated. They apparently contribute significantly to geomagnetic variations on Earth, and they probably cause further heating of the solar wind even after it has passed the Earth's orbit.

What is the ultimate fate of the solar wind? Does it continue indefinitely into space as a supersonic flow or does something stop it?

As the solar wind moves away from the Sun, so long as there are no physical forces to slow it down, its concentration must fall off according to an inverse square law. Consequently, the dynamic pressure that it can generate by flowing against any obstacle decreases with increasing distance from the Sun. It is reasonable to expect, therefore, that it will finally become so feeble that it will not be able to push aside the Galactic magnetic field and make a space for itself. From radio measurements, we have good indications that there is a Galactic magnetic field whose intensity is about one gamma (a gamma is 10^{-5} gauss).

When the solar wind becomes too weak to push aside the Galactic magnetic field, its supersonic flow will be stopped, and a shock wave must be expected to form. Beyond the shock wave, the gas particles will flow subsonically, and most of the energy previously associated with the supersonic flow will be present in the form of thermal motion. In other words, the gas is heated by passage through the shock front. In addition, the gas is compressed on passage through the shock front, and the magnetic field embedded in the gas is also compressed. It should be noted, however, that the total particle energy (in both ordered and thermal motion) far exceeds the magnetic energy.

The concept that the solar wind passes through a shock front only replaces the supersonic outflow with a subsonic outflow: the outflow of solar gases does continue, though at a slower rate, still dragging the magnetic field along with it. What probably happens to finally stop this combined motion of magnetic field and gas is that the gas cools and reduces its conductivity so much that the magnetic field can slip through the gas. The magnetic field includes oppositely directed components which annihilate one another as they slip through the plasma, the magnetic energy being transferred to the plasma.

How the plasma cools is interesting. Galactic space contains cold atomic hydrogen, whose pressure and temperature have been determined by means of radio signals emitted by the hydrogen atoms. This cool atomic hydrogen drifts into the region occupied by the hot plasma behind the shock front. When a hydrogen atom collides with a rapidly moving hydrogen ion, there may be a charge transfer process in which the ion is neutralized and the atom becomes an ion; the effect is just the same as if the energies were exchanged between the atom and the ion. In this way, the hot ions behind the shock front transfer their energy to neutral hydrogen atoms, thus cooling the plasma. After the plasma has cooled, its electrical conductivity is low enough so that the magnetic field lines can slip through the plasma in a reasonably short time.

On the basis of evidence that I shall discuss below, the shock front appears to be located at a distance of about 20 astronomical units from the sun, near the orbit of Uranus, one of the outermost planets of the solar system. The concentration of hydrogen ions in the solar wind just before reaching the shock front is only about one hundredth of an ion per cubic centimeter. On passing through the shock front, the concentrations are increased further in order to bear the pressure between the shock front and the galactic magnetic field.

As the magnetic fields within the cool plasma merge and annihilate one another, the cold plasma is released from its last bond with the solar system, and gas clouds can drift out into Galactic space unencumbered by the magnetic field. The thickness of the zone of cooling and annihilation of magnetic energy is probably between 5 and 10 astronomical units. Beyond this—that is, beyond a distance of about 30 astronomical units from the Sun—one is outside the solar system from a gas-dynamics point of view.

I mentioned earlier that cold atomic hydrogen exists in galactic space. Radio measurements indicate that the concentration is about 1 atom per cubic centimeter and that the temperature is very low, less than 100 degrees Kelvin. What we would like to know is how much and how far this extensive cloud of hydrogen atoms permeates the solar system. There are two processes that will tend to keep it out. The first is photo-ionization. As the hydrogen atoms drift into the solar system, they will be exposed to greater and greater intensities of solar ultraviolet radiation capable of ionizing the atoms. Once ionized, the atoms are lost as atoms, because recombination proceeds much too slowly to replace them. The ions produced by photo-ionization will be swept up by the solar wind and blown to the outer parts of the solar system as part of the solar wind.

The second process that can keep hydrogen atoms out of the solar system is charge exchange with the solar wind. Whenever such a charge exchange occurs between a galactic atom that has drifted into the solar system and a hydrogen ion in the solar wind, the hydrogen atom acquires the solar wind velocity and is rapidly removed from the solar system. As a consequence of these two processes, the cold galactic hydrogen does not penetrate deeply into the solar system, although it does penetrate the outer portion. If these were the only ways in which neutral hydrogen could enter the solar system, one would expect very small concentrations within a few astronomical units of the Sun.

To get significant numbers of hydrogen atoms relatively near the Sun, it is necessary to send them into the solar system with much greater velocities than those that are appropriate to cold galactic hydrogen, so that they can approach the Sun rapidly and get fairly close before they are lost by ionization or driven back by charge exchange with the solar wind. And there is a source of rapidly moving hydrogen atoms of just the sort that is required. This source lies just beyond the shock front at the outer limits of the solar system. The source is the charge exchange that occurs between cold galactic hydrogen and hot hydrogen ions that were given high thermal velocities by passage through the shock front. Charge exchange not only cools the ions in the plasma, but it transfers energy to hydrogen atoms that then fly off in all directions, some of them directed toward the Sun.

Rapidly moving hydrogen atoms that arise beyond the shock front can penetrate rather deeply into the solar system before they are lost by ionization or charge exchange simply because their velocities are high enough. They can get fairly close before there is time for the loss processes to be effective. In fact, very little loss occurs beyond the orbit of Jupiter, about 5 astronomical units from the Sun. The loss becomes rapid inside the orbit of Mars—about 1.5 astronomical units—and very little of the hydrogen can penetrate within the Earth's orbit without being lost.

The concentration of neutral hydrogen within the solar system is thus dependent upon the location of the shock front: the closer it is to the Sun, the more hydrogen it will provide within the solar system. This is so because the amount of hot hydrogen leaving the region beyond the shock front, going in all directions, must equal the amount of ionized hydrogen entering the region. The closer to the Sun the shock front, the larger will be the influx of solar wind ions per unit area, and the larger will be the outflow per unit area of hydrogen atoms. And a larger outflow per unit area of hydrogen atoms will lead to larger concentrations of neutral hydrogen within the solar system. Therefore, if we can determine the concentration of hydrogen atoms in the solar system, we can determine the distance to the shock front.

Some measurements exist that do permit the evaluation of the amount of neutral hydrogen in the solar system and hence the distance to the shock front. These measurements are of Lyman alpha radiation from hydrogen, the radiation that results when the electron of the hydrogen atom moves from its first excited level to the ground level. The Sun is a powerful emitter of such radiation; its Lyman alpha line is the most prominent of all line emissions in the solar spectrum. Because of the high temperature on the Sun and other complicating factors, the solar line is a rather broad one. When this radiation falls on the Earth's atmosphere, some of it is resonantly scattered by atomic hydrogen that constitutes the outermost portion of the Earth's atmosphere. Because the Earth's atmosphere is relatively cool, the telluric Lyman alpha line, which appears as an absorption line in the middle of the solar emission line, is very narrow compared to the solar line, and only a narrow core is absorbed from the center of the solar line. The absorbed radiation is re-emitted as fluorescence radiation, and the fluorescent line is narrow, not wide like the solar line. At night, this radiation can be seen by instruments flown in rockets, and such observations provided the first indication that the Earth's outermost atmosphere consists mainly of atomic hydrogen. This led to the concept of a geocorona, or telluric hydrogen corona.

The detector that is used to observe the nighttime Lyman alpha radiation can be made insensitive to the radiation arising in the geocorona by placing over it a filter containing some atomic hydrogen. If the appropriate amount of atomic hydrogen is placed in the filter cell, it can be made to absorb completely over a wavelength interval that is broader than the fluorescent line. Then, if the nighttime radiation comes only from the geocorona, there should be no response. When D. C. Morton and J. D. Purcell performed this experiment, they found that some response remained. The response is to be identified with the resonance scattering of solar Lyman alpha radiation by hydrogen atoms in interplanetary space. This measurement thus provides the means of determining the neutral hydrogen concentration in interplanetary space, for the intensity of fluorescence is dependent upon the concentration. The answer is that the concentration is about 0.03 atom per cubic centimeter beyond 5 astronomical units, falling to 0.01 atom per cubic centimeter near the Earth's orbit, and much smaller values at smaller distances from the Sun. The shock front is able to provide these hydrogen concentrations if it is located at a distance of about 20 astronomical units from the Sun.

I have just given a description of the solar system from a gas-dynamics point of view that could be summed up in the following way. The solar corona is the source of the solar wind that blows through interplanetary space with supersonic velocity. At a distance from the Sun of about 20 astronomical units, it loses its supersonic characteristic on passing through a shock wave. Beyond this, it cools and finally drifts off into galactic space as clouds of cool gas. The outer limit of the solar system, in this view, is about the same as the outer limit of known planetary orbits.

The region beyond the shock front serves as a source of rapidly moving, or hot, hydrogen atoms, and some of these penetrate deeply into the solar system. This neutral hydrogen within the solar system constitutes a very tenuous gas, but it does scatter solar Lyman alpha radiation to a perceptible degree. It is possible to make much more sophisticated measurements of the scattered radiation than have been made in the past, and it is to be anticipated that these measurements will soon be made. The results of these measurements, when they are available, will contribute to our increased knowledge and understanding of the solar system, particularly with respect to the forces and processes involved in interactions between fields and charged particles.

5

Sun and Earth

E. N. PARKER

The main outpouring of energy from the Sun is its light. The light supplies most of the warmth that the Earth receives from the Sun. The changing of warmth from day to night, summer to winter, and sea to land is responsible for much of our weather at the surface of the Earth. Interestingly enough, there are other forms of energy from the Sun which, though smaller in quantity than the light, produce weather at higher levels in the atmosphere and are extremely important, each in its own locality. Let me cite a few examples.

Ultraviolet light from the Sun, which is similar to visible light except that its wavelength is shorter, excites the higher regions of the atmosphere and produces the upper levels known collectively as the "ionosphere," which has so much effect on radio waves and their propagation around the Earth. The ionosphere is the outer atmosphere of the Earth, where a significant portion of the atoms and molecules have lost an electron. The ionosphere starts at an altitude of about 100 kilometers and extends outward rather indefinitely for enormous distances into space.

Solar X-rays are another example. The quiet Sun is a weak source of X-rays, but when there are large sunspots and increased solar activity, X-ray emission from the Sun can become intense. At such times both the ultraviolet and the X-rays are enhanced and rather profoundly alter the ionosphere of the Earth for periods of hours or days.

If we look toward the long wavelengths there is, of course, the infrared, which contributes mainly warmth to the Earth, though less than does visible light. The infrared contributes warmth at high altitudes in the terrestrial atmosphere, in contrast to the visible light, and thereby plays its own role in weather. At even longer wavelengths, there are radio waves from the Sun. The quiet Sun is a weak emitter of radio waves, so it can be detected only with suitably sensitive radio equipment. The active Sun sometimes emits enormous bursts of radio noise which are not only relatively easily detected, but also tell us about things which go on at the Sun.

Altogether, then, there is a very broad spectrum of electromagnetic waves, involving X-rays, ultraviolet, visible, and infrared light, and radio waves, which come to the Earth from the Sun. The visible light is the main source of energy, but each wave length outside the visible produces its own individual effects in its own region.

Entirely different from the electromagnetic waves which I have spoken of so far is the corpuscular radiation from the Sun. Particles of matter—electrons, protons, helium nuclei, etc.—are emitted from the Sun at various energies, at various times, and each has its own interesting and exotic effects. Corpuscular radiation has not been known for as many years as the electromagnetic radiation because its detection requires advanced technological skills. Usually one has to get high in the atmosphere before he can detect it at all. The most spectacular corpuscular emissions occur at the time of solar flares.

A flare, described in chapter 3, is a sudden brightening of the outer atmosphere of the Sun in the vicinity of large sunspots. It is not known just what a flare is except that it is an explosive storm on the Sun, involving rather large amounts of energy. At such times particles, particularly electrons and protons, are produced and stream out into space along the magnetic fields of the Sun with velocities sometimes approaching very closely the speed of light. It is these fast, or relativistic, electrons which cause many of the radio bursts from the Sun at the time of flares. The protons which are emitted at such times escape from the Sun into interplanetary space and fill large volumes of space with intense fluxes of fast particles. (They are, of course, a hazard to any astronaut who may be out in space at that time, and a good deal of thought has been put into this problem, in view of the plans for journeys into space.) These particles come to the Earth where, for some twenty years or more, they have been detected and intensively studied.

At this point, let us turn to the general nature of the Earth's magnetic field. The magnetic field comes out at the south pole of the Earth, swings around through space, and goes in at the north pole, so that over most of the Earth we have a thick region of magnetic field over our heads. In the polar regions, where the lines of force are coming straight in from space, it is possible for fast particles from the Sun to come in along the lines of the magnetic field. They do not travel easily across the lines of the field. In the polar regions, then, the fast particles from the Sun come in to the Earth and profoundly affect the ionosphere. At such times, when the Sun is extremely active and the ultraviolet, X-ray, and fast-particle emission from the Sun is extremely high, the ionosphere may be seriously disrupted and irregular, with the result that radio communications are broken or extremely difficult for extended periods of time.

In addition to the bursts of fast particles from solar flares, there is a somewhat more general, broader emission from the Sun which goes on all the time. I am referring now to the relatively low-energy and slow emission of matter from the Sun that is responsible for magnetic storms, trapped radiation belts, comet-tail behavior, and the aurora to name a few of the effects. Historically, the magnetic storm and the aurora have been known for some time, the aurora, probably, for hundreds of thousands of years. The magnetic storm was detected as soon as men learned to make compass needles, which was some four or five hundred years ago. A few sharp-eyed and patient observers found that the needle, which normally hung steadily, was occasionally disturbed. And after a period of decades, it was discovered that this occurred particularly at the time of high auroral activity. The origin of the magnetic storm and the aurora was traced to the Sun, but the origin and nature of the corpuscular emission from the Sun responsible for them remained a mystery until modern times. Curiously enough, comets provided the explanation. It was observed that gaseous comet tails point away from the Sun, and we finally concluded that this was caused by the rush of corpuscular emission past the comet. The fact that the comet tails always point away from the Sun showed that the corpuscular emission was a very general phenomenon, that it did not require special conditions on the Sun, and that its origin must lie in some general property of the Sun and its outer atmosphere.

To understand how the general emission of low-energy corpuscles from the Sun takes place, we should digress for a moment and consider the Sun itself. The center of the Sun is extremely hot and extremely dense. The temperature is 15 million degrees, approximately, and the density is approximately one hundred times that of water. The material is a gas, in spite of its high density, because it is so hot. It is in the center of the Sun that the energy supplying the Sun is released from the nuclei of hydrogen atoms through conversion into helium. The energy is carried outward through the Sun by the radiation buried in the Sun. X-rays predominate at the center of the Sun, and it is by their slow diffusion and transfer outward, with steadily dropping temperatures, that the energy finally reaches the 6,000-degree surface of the Sun. The outer layers of the Sun behave like the water in a kettle under which a fire has been built. The under side of the layers is much hotter than the upper side of the layers, with the result that the layers boil, or convect, as one says. They constantly turn over. This convection can be observed on the surface of the Sun. The surface of the Sun, if you look at it with suitable instruments so that it does not dazzle the eye, can be seen to be a large, boiling pot. Above the visible surface of the Sun the atmosphere extends out for a considerable distance, and this corona can be observed for 10 million kilometers into space. The interesting thing is that the boiling at, and beneath, the visible surface of the Sun generates so much agitation—sound waves, gravity waves—that the tenuous outer part of the atmosphere is violently heated. The temperature rises once more to millions of degrees—in this case, 2 or 3 million degrees—to form the solar corona. The corona (Latin for "crown") is so named because it is observed as a white halo around the Sun, extending far into space, when the bright disk of the sun is cut off by the solid body of the Moon during a solar eclipse. The outer atmosphere of the

Sun is extended into space for long distances by its enormous temperature of 2 or 3 million degrees. It is this outer atmosphere of the Sun which is responsible for the continual emission of particles from the Sun that produce the magnetic activity and the aurora.

It turns out that an atmosphere as hot as the atmosphere of the Sun, extending far into space with a temperature of a couple of million degrees, has no static equilibrium but must continually flow into space. There is no way of shutting off the flow, short of building a large box around the Sun. The continual expansion of the outer atmosphere of the Sun leads to winds blowing from the Sun at the rate of 300 to 800 kilometers per second. These winds make the journey from the Sun to the Earth in a period of 2 to 5 days and have been called the solar wind. It is the flow of the solar wind around the Earth, and particularly over the outer boundaries of the magnetic fields of the Earth, which produces the magnetic and auroral effects with which we are familiar.

I mentioned earlier that the Earth has a magnetic field which extends out into space and shields the Earth from fast particles from the Sun. Even more effectively, it shields the Earth from the solar wind itself. One must go some 60,000 or 70,000 kilometers into space in the direction of the Sun to get out of the magnetic field of the Earth. Were it not for the solar wind, the magnetic field of the Earth would, of course, extend much further into space. But the effect of the wind is to compress the field around the Earth and to confine the field within a limited region called the magnetosphere. The magnetosphere is a comet-shaped region terminating on the sunward side of the Earth at a distance of some 60,000 or 70,000 kilometers but extending off in the antisolar direction for distances which must be at least a million kilometers (no one knows yet exactly how far). The boundary of the magnetic field appears to be rather well defined. As far as observations can tell, the thickness of the boundary is perhaps no more than 50 or 100 kilometers. The boundary is a transition from the magnetic field and a very tenuous outer atmosphere of the Earth to the rapidly streaming solar wind coming to us from the Sun.

The flow of the wind over the boundary of the magnetosphere is responsible, apparently, for magnetic activity and for the aurora. Magnetic activity consists of a general shaking and fluctuation of the magnetic field. The violent activity has very little pattern.

The magnetic storm starts when a particularly strong blast of wind comes from the Sun and compresses the magnetic field in much closer to Earth than normally. This is accompanied by an increase in the field strength in the vicinity of Earth. It is a small increase, a fraction of a percent, generally. But it can be easily detected with suitable instruments. A little later, when the storm has progressed in the usual way, gases from the solar wind apparently fold into the magnetic field, inflating the magnetic field and causing it to spring outward into space. This is the opposite from the way the storm begins. It is called the main phase of the storm and may go on for several hours or days.

While the solar wind is compressing and then expanding the magnetic field of Earth, particles apparently are folded in to the magnetic lines of force, both in the long tail extending out behind the Earth and perhaps elsewhere around the boundary. The magnetic field is, at the same time, convecting slowly under the forces exerted on it by the solar wind, so the particles are carried deep into the field. The particles are accelerated by perhaps several mechanisms and, in their journey in the rolling magnetic field, find themselves in collision with the atmosphere of the Earth. This is apparently the basis for the aurora, which has been recognized for seventy years now as the result of fast particles coming into the atmosphere of Earth. The particles that produce the aurora are different from the very fast particles from the solar flares, which I mentioned earlier. It is true that the fast particles from the solar flares may contribute to some of the luminous emission, but apparently not to the extremely luminous displays that we call the visible aurora.

Let me digress for a moment to discuss the acceleration of particles. Up to the present point I have stated merely that the solar wind blowing over the magnetosphere of the Earth causes particles to be accelerated by one or more of several mechanisms. The mechanisms are not all clearly understood in each separate application. We should recognize that the acceleration of particles from ordinary thermal speeds (of a few kilometers per second) to extremely high velocities, sometimes reaching near the speed of light, occurs in many circumstances throughout the universe. The general rule seems to be that whenever gases move rapidly in the presence of magnetic fields, they produce a few particles with extremely high velocities. There have been several ideas suggested to explain the means by which the high velocities are achieved. One process for acceleration involves the jostling of particles back and forth between massive elements of gas and field. Imagine a light object bouncing repeatedly from massive moving elastic objects and with each bounce achieving a slightly higher velocity on the rebound. This random process of acceleration is called the Fermi mechanism and is believed to be widely operative. It is by no means the only possibility, however. There are several other schemes that have been thought of.

One of the most prominent of such schemes involves particles moving around in a circle in a magnetic field. This circular motion is sometimes called cyclotron motion, in analogy to the laboratory instrument in which the technique is used. As the particles move around the magnetic field, a periodic push is given to the particles once each time around, due to some passing wave. Thus the particles go faster and faster, gaining a small amount of energy each time around. This is the basis for the design of the laboratory cyclotron. It may also work in space, where one has waves of all frequencies present and therefore some waves with the right frequency to accelerate particles.

Finally, perhaps the simplest acceleration scheme of all, one that has been known for decades, is *adiabatic compression*. This phrase simply means that when you compress a gas, such as air, you do work on it, and the gas gets hot. Anyone who has pumped up a bicycle tire or an automobile tire with a hand pump knows how hot gas can get from being squeezed. Gases in space, and the magnetic fields that pass through the gases, may be compressed under a variety of circumstances. In connection with the aurora, probably the most interesting compression takes place when the magnetic field of the Earth overturns slowly and lines of force which were initially far out in the magnetic field and widely separated are brought in close to the Earth and squeezed close together. The compression in this case may be as much as by a factor of a thousand, or even ten thousand in extreme cases. Roughly speaking, the energy of a particle increases by about the same factor as the volume which it occupies decreases.

Individual electrons and protons—that is to say, fragments of atoms such as are found in the solar wind—may be picked up in the magnetic field of the Earth out near the boundary of the field, where the strength is perhaps one ten-thousandth of a gauss, and carried in along with the magnetic field (in the course of several hours) to be compressed into a region where the field has a strength of a half of a gauss. The compression is then by a factor of more than a thousand; and particles that initially had temperatures of the order of 10,000 or 20,000 degrees, which is very modest for the solar wind, would then find themselves with temperatures of the order of 20 or 50 or 100 million degrees. In this way the compression may be responsible for many of the fast particles which produce the aurora. The aurora, of course, is an extremely complicated phenomenon and there must be more to it than this. But this may be the basis for it.

Now let us look at the broad problem of acceleration of particles in the universe. The general rule is that violent gas motions and fields lead to fast particles. These fast particles may produce a number of effects such as the aurora, or in other cases (in stars) they may produce intense radio emission. They may be responsible for much of the phenomenon called the solar flare on the Sun. What I want to emphasize, however, is that the aurora, which is so spectacular on the planet Earth, is only a special case of a very widely occurring phenomenon in the universe, agitated gases producing fast particles.

Observations on the solar wind outside the magnetic field of Earth show that there are large numbers of fast particles there too. Some of these come from solar flares on the Sun; some of them come from the Galaxy and are called cosmic rays. But some of them seem to be produced locally, not inside the magnetic field as are the auroral particles, but as the wind strikes violently against the magnetic field, goes through a shock transition, and flows with some turbulence around the sides of the magnetic field. It is to be hoped that in the next few years increasing observational information will help to disentangle the general storm of particles and perhaps lead to a somewhat more detailed understanding of their origins than is presently possible.

Historically, the first major concern with the acceleration of particles in nature came from inquiries into the origin of cosmic rays. Cosmic rays have been known since the 1930's to consist of fast protons plus other particles, and speculation arose as to the origin of these fast particles. As observations extended into other fields—and, in particular, into radio astronomy—we recognized that throughout the Galaxy there were many regions where intense fluxes of fast particles occurred. Thus the search for the origin of cosmic rays has broadened greatly, until now one associates the whole problem of particle acceleration with supernovae, solar flares, aurora, and so on.

Coming back now to the question of the solar wind and the Earth, I want to note that, in addition to auroral displays and magnetic storms, which are most common when the Sun is active, there is some evidence that solar activity has influence on the ordinary terrestrial weather. The influence is not of a simple nature. It is not that rain comes when the Sun is active or fails when the Sun is active, but rather that the overall terrestrial weather pattern seems to be somewhat different when the Sun is active than when the Sun is not active. When the Sun is not active, the wind patterns in the northern and southern hemispheres of the Earth tend to settle down into strong east-west circulations, whereas in times of solar activity there is a greater tendency for the formation of individual storms which give strong north-south mixing. This is of course a statistical matter. There are always storms associated with north-south mixing, even when the Sun is inactive. When the Sun is active, there seems to be a stronger tendency for the north-south circulation as compared to years when the Sun is inactive. It is the statistical nature of the effect which leads to the possibility for controversy about the reality of the effect.

Without going too far into speculation, we may nevertheless grasp the important implications of solar activity by applying the idea as an explanation for the extensive ice ages of the past. People have asked for many years: How is it that sometimes the Earth can be warm, with a semitropical climate extending almost to the poles, and at other times have polar conditions extending down to the middle latitudes? It is not a simple matter of the Sun being hot or cold, because the indications are that the Sun has shone with its present brightness for billions of years. Presumably, it must be some subtle change in conditions in the terrestrial atmosphere. One can see that if the atmosphere were to settle into a strong east-west circulation for an extended period of time, there might be the possibility that the warm air from tropical regions would not mix up into the middle latitudes, and therefore polar conditions could be obtained. This is of course a speculation. The connection of terrestrial weather patterns with activity on the Sun may be eventually a means for understanding at least some part of the question of the formation of ice ages on this planet. Observations of solar activity show that, in fact, it is extremely irregular and there have been periods of some twenty or thirty years when the Sun was essentially without much solar activity. This is not long enough for ice age conditions to develop, certainly. But it shows that solar activity, which is itself not really very well understood, may be something that can shut itself off for extended periods of time, perhaps thousands of years. We simply do not know.

What about immediate effects of solar activity on the terrestrial weather pattern? The mechanism by which this effect takes place is not well understood. Indeed, the existence of such influence is stoutly denied by some experts in the field. But in spite of such opposition, there seems to be some relationship, and there are some ideas as to its basis. For instance, one idea has to do with the possible role of ions in rain formation. When there are large numbers of aurora, large numbers of ions are produced in the upper atmosphere. Each such ion is a seed for condensing moisture; we are all familiar with the effects of seeding clouds with small crystals of various chemicals. The crystals enhance the condensation of moisture, leading to cloud formation. Presumably, seeding high regions of the atmosphere with ions should enhance the condensation of ice crystals, which, in turn, give an increased "greenhouse" effect[1] and in that way may have some profound effects on the weather. This is not proved, but it looks plausible.

Present worldwide space programs have contributed a great deal to understanding the effects which the corpuscular emission from the Sun has upon the Earth and, for that matter, to understanding the ultraviolet, X-ray, and infrared emissions from the Sun. The problem is that these emissions from the Sun cannot be observed at the bottom of the atmosphere. We are shielded

---

[1]Ths effect is so-called because it is commonly experienced in greenhouses. Even on a very cold day light passes through the windows of a greenhouse. Striking interior surfaces, plants, and soil, the light radiation is re-radiated at the longer, infrared wavelengths. These waves, however, cannot pass out through the window: while glass readily transmits light waves, it blocks the lower frequency infrared waves. Thus the greenhouse warms up as more and more light radiation enters. Moisture and carbon dioxide in the atmosphere serve as similar barriers to re-radiated energy from the Earth.

from them by the dense cloud of air over our heads. One could argue that it is a very good thing we are shielded, because most of these radiations are bad for life of any form that we know; nonetheless, it has been possible only in the last ten or fifteen years to begin to get up out of the atmosphere with sufficient ease to make meaningful measurements. Once begun, however, progress in studying and understanding these phenomena has been very rapid. Ten years ago an essay on this same subject would necessarily have been far more sketchy, and the statements far more dubious. Ten years ago we did not understand clearly the origin of the corpuscular radiation producing magnetic storms. A number of ideas had been constructed, but they were vague and controversial. Extremely fast particles from flares had been detected at that time, but again the information available was not sufficient to come to any clear understanding of how they moved through space and exactly what effects they have on the Earth.

This is not to say that things are entirely understood at the present time. You will have noted a number of vague statements in the remarks that I have made. For instance, in discussing the origin of magnetic storms and the aurora, I have asserted that they are caused by the solar wind blowing over the magnetosphere of the Earth. I think that is correct. But it is very difficult to say exactly how this effect takes place. There are many ideas as to this, and I have my own opinions, of course. I think the basic requirements for the aurora and magnetic storms are fairly well understood now, but whether the radiation belts, the aurora, and magnetic storms are produced mainly by convection of the magnetic field of the Earth, or by acceleration of particles in the neutral sheet in the magnetic tail of the Earth, is still a subject of considerable debate. We can see that, with continuing space observations, the question should be subject to eventual settlement in principle.

II

THE EARTH AND THE MOON

• 6

The Core and Mantle

ANTON L. HALES

For many years it was thought that the Earth was liquid below a comparatively thin crust. During this century, seismology, the study of the travel times of the waves from earthquakes, has led to a picture of the Earth's interior which is definite in broad outline, but which leaves open a number of intriguing questions which I shall discuss later. It is by probing these questions that we can hope to gain understanding of the forces which have shaped the Earth as we see it today, the forces which create mountains and conceivably move the continents.

Let us start with the definite and then turn to the interesting areas of uncertainty. We are now convinced that the solid Earth can be viewed as consisting of three regions: the very thin crust, the mantle, and the core. The crust is about 30 to 40 kilometers thick in continental areas and only about 6 kilometers thick under the oceans. Beneath this surface layer lies the mantle, which extends to a depth of about 3,000 kilometers. The mantle is divided conventionally into an upper part and a lower part. The upper mantle is the outermost part, between 400 and 1,000 kilometers. Here we know that the material changes in composition with depth, and recent studies have shown significant differences between the region lying beneath the oceans and that beneath the continents. The mantle is solid except possibly in localized regions—magma chambers—where the lava of the volcanoes is generated before being spewed out at the surface.

Beneath the mantle, in turn, lies the core, and this region, ranging from the Earth's center out to some 3,000 to 3,400 kilometers, consists of two distinct zones, the inner and outer cores. The inner core is solid and has a radius of 1,200 to 1,300 kilometers, while the outer core, extending out to perhaps 3,400 kilometers, is liquid.

These gross structural features, as well as almost every detail we have about the deep Earth, are revealed by seismic waves arising from earthquakes. However, there are two kinds of seismic waves in a solid, P waves and S waves. P waves have a longitudinal motion, that is, the particles move in the direction of propagation of the waves. S waves have a motion that is transverse to the direction of propagation. One may think of these as "push" and "shake" waves. The P waves travel at speeds of 8 to 13 kilometers per second in the mantle, while the slower S waves have speeds of about 5 to 8 kilometers per second. One other difference is most important: only P waves can travel through a liquid medium. At any interface or discontinuity in elastic properties, it is possible to generate refracted and reflected P waves as well as refracted and reflected S waves. This transformation of the seismic waves into the other kinds of waves at an interface adds considerable complication to the study of records of earthquakes at great distances, but it is, nevertheless, fortunate because it provides much additional information on the structure of the Earth.

At the boundary between the core and mantle, for example, an incident P wave is reflected, but part of the same P wave travels on into the core and finally back from the core out to the surface. But we cannot find any evidence of an S wave having traveled through the outer core, and this is one of the reasons for believing the outer core to be liquid. Instead, that part of the S wave which is not reflected travels through the core as a P wave. Because of the difference between the S and P velocities outside the core, the paths within the core are rather different, and thus we have two sources of information on velocities in the core. It is the comparison of the seismic velocities (determined from the travel times of the seismic waves) with the results of laboratory experiments at high pressure and temperature which enables us to make intelligent estimates of the materials of the Earth at great depth.

The pressures in the core, greater than 1.4 million atmospheres, are far beyond the range of laboratory high-pressure apparatus, and so we rely on shock-wave studies in which an explosive charge generates a shock wave that exposes samples for a few microseconds to pressures as great as those existing at the center of the Earth. These studies lead to the view that the core is made of iron, possibly alloyed with a few percent of some lighter element such as silicon.

If, then, the core is made of iron, one can make inferences with regard to the temperature of the core, using extrapolation of the melting-point curve for iron measured at lower pressures. These extrapolations rest on the application of solid-state physics theory in one form or the other, and lead us to the conclusion that the temperature at the core mantle boundary is between 4,000 and 5,000 degrees centigrade.

Here I would like to call attention to an important question. This has to do with the energy required to drive the convection currents in the liquid outer core. These currents, it is now thought, are responsible for the magnetic field of the Earth. There appear to be only two possibilities that can account for this energy. One is that a significant portion of the radioactive material of the Earth is in the core. The other is that the solid inner core is growing and releases heat to the outer core as it solidifies.

So far, I have made use of information about the Earth which was derived from the travel times of the P and S body waves, the waves which travel through the Earth like light waves through a lens. But the record of an earthquake contains other information, and here I am referring particularly to waves traveling near and at the surface. There are two kinds of surface waves, Rayleigh and Love waves. These waves travel round the surface of the Earth rather than through it. They have long periods, up to several hundred seconds, and so, although they are traveling round the Earth, the velocity with which they travel is affected by the properties of the material at depths of several hundred kilometers. These waves have provided valuable information on the properties of the upper mantle. In fact, one might say that it was the analysis of these waves, and the inference from this analysis that there was indeed a Gutenberg low-velocity layer in the upper mantle, which led to the increased interest in the upper mantle and which has culminated in the international Upper Mantle Program. Modeled much along the pattern of the International Geophysical Year, this program began in 1962 and runs to 1970. It is a coordinated attack by scientists of some fifty nations on the problems of the upper mantle and the crust.

Closely related to the study of surface waves is the study of the free oscillations of the Earth. It turns out that the Earth rings like a bell for many days after large earthquakes such as the recent Alaskan and Chilean quakes. These oscillations are of two kinds, radial and toroidal, and have many modes, ranging in period from about 55 minutes down to the periods of hundreds of seconds in the surface waves. As might be expected, the periods of the low-frequency modes depend more on the properties of the material at great depth, and the periods of the higher frequency more on the material in the outermost few hundred kilometers.

These free oscillations are beautifully recorded on the strain seismometers developed by Hugo Benioff at the California Institute of Technology. These instruments measure how much 100 meters of the Earth stretches when a seismic wave passes by. It was Benioff who first realized that it was possible that the long-period waves after a major earthquake could be the free oscillations of the Earth. Some of the modes of free oscillations are also very well recorded by the tidal gravimeters developed for the International Geophysical Year program.

Studies of P and S waves and of Rayleigh and Love surface waves have provided us with a model of velocities within the Earth. From this model the periods of the free oscillations could be calculated, permitting us to compare the calculated oscillations with those derived by analysis of the strain and tidal gravimeter records. Thus we would have a comparison of results based on two independent sources of information, one from the seismic body waves, the other from the periods of the free oscillation. But the calculation of the periods of the free oscillations required the variation of density with depth. How the Earth material varies in density is not only significant to this comparison but is clearly important in itself because it tells us much about composition and structure.

This variation of density with depth had been studied by K. E. Bullen of Australia. The density in any chemically homogeneous portion of the Earth varies with depth in a manner given by the famous Adams-Williamson equation. Bullen integrated the equation and showed that, since the calculated mass and moment of inertia of the model had to be consistent with the observed mass and moment of inertia, the whole mantle could not be chemically homogeneous; there was necessarily either a density discontinuity or a region of rapid change of composition at some depth between 300 and 900 kilometers, that is, in the upper mantle. It was difficult to determine how the density varied in this region, and Bullen over the years developed two models which have come to be regarded as reasonably good representations of the density within the Earth.

When the periods of the free oscillations were calculated on the basis of the body-wave- and surface-wave-velocity models and the Bullen density model, it was found that there was, in general, excellent agreement between the calculated and observed periods. The exceptions were for the low-frequency radial and toroidal modes, the periods of which depend in large measure on the properties of the Earth at depths between 1,500 and 3,000 kilometers. Several suggestions have been made for changes in the model to remove this discrepancy. All involve some modification of our present models of the lower mantle and core, the density, the velocities, or the radius of the core. This is one of the areas of uncertainty which is being studied very closely at the present time.

Now let us turn to the upper mantle. It is, I think, commonly believed today that it is in the understanding of the properties of this region that we shall find the clues to the Earth-shaping processes.

It was in the upper mantle that the Gutenberg and Jeffreys-Bullen velocity distributions differed. The Gutenberg distribution had a low-velocity zone in the upper mantle; the Jeffreys-Bullen did not. One of the first results of the investigation of the surface-wave velocities was to show that for S waves there was a low-velocity zone, as had been stated by Beno Gutenberg. The zone was closer to the surface and probably more marked below the oceans than below the continents. Later work has shown that there are differences in the S-velocity structure between the ancient shield areas and the regions of more recent tectonic activity, a good example being the differences between the regions to the west and east of the Rocky Mountain front.

Thus we see that the differences in structure which are seen at the surface do not, as had once been thought, stop at the Mohorovičić discontinuity, the boundary between the crust and the mantle, but extend to depths of several hundred kilometers. Recent studies of the travel times from large explosions at distances of between 1,000 and 3,000 kilometers have confirmed that there are regional differences in the sub-Moho velocities. More surprising, new studies of the travel times of the seismic waves at teleseismic distances (3,000 to 11,000 kilometers) have shown that there are significant regional anomalies of the order of one or two seconds. These are interpreted as implying differences in P velocity extending to depths of several hundred kilometers. Clearly, it is necessary to map the upper mantle in much the same way as geologists map the surface. The detail will not be as great and the techniques will be different and less direct, but the results may be just as important for our understanding of the history of the Earth. This mapping of the upper mantle is undoubtedly one of the keys to the understanding of the Earth-shaping processes.

It has long been recognized that the attenuation of seismic waves depends on the physical properties of the material at depth, but in the case of body waves it is difficult to separate the geometrical focusing effects of regions of rapid velocity change from those of absorption. Recently, methods of studying the anelasticity of the Earth by the use of surface waves have been developed. The results are expressed in terms of the proportion of the energy lost by a wave during one cycle. This quantity is high, about 1/100 to 1/200, for the low-velocity zone, and surprisingly low, of order 1/2,000, for the lower mantle. One can interpret these results as implying much lower viscosities in the low-velocity zone than in the crust, or lower mantle. The inference of low viscosity in the low-velocity zone is confirmed by recent studies of the Fennoscandia uplift.

What does this mean? Broadly speaking, geophysicists had expected that the Earth would be hotter closer to the center and that therefore the center would be closer to the transition between solid and liquid. Thus, the Earth would be weaker near the bottom of the mantle than near the top. But these new studies of the anelasticity show that the weakest zone is in the upper mantle, that is, in the outermost 200 kilometers. It is here that the movements of the continents relative to one another must take place. It is because the lower parts of the Earth are so strong, more like steel than like putty, that the Earth rings like a bell after a large earthquake and that sizable movements of the ground, up to a half-inch, go on in Washington, D. C., for long after large earthquakes such as those in Chile or Alaska.

The discussion thus far has been confined to information on the Earth's interior derived from seismic studies. There are, however, other geophysical and geochemical data which bear on the composition and structure of the interior. It is known that, by and large, the Earth is in a state of isostatic equilibrium: in other words, the mass per unit cross section is roughly the same in any vertical column whether it is below the continents, the oceans, the high mountains, or the sea-level plains. Since the rocks of the crust are known to be lighter than those of the mantle and are known also to have been derived from the mantle, there is, as was pointed out by the young American geophysicist G. J. F. Macdonald, good reason for believing that on the average there has been little horizontal movement of the crust relative to the upper mantle. Support for this conclusion comes from the studies of the heat flow. Since Sir Edward Bullard developed, about 1948, a method for measuring heat flow at sea, many measurements of heat flow have been made at sea. It is clear that the heat flow through the ocean bottom averages between 1 and 1.5 microcalories per square centimeter per second, just as it does on land. It is known that about half of the heat flow on land comes from the radioactivity of the rocks of the crust, so that again one is driven to the conclusion that the mantle below the continents has stayed there throughout the history of the Earth. Thus, as has been pointed out by Francis Birch at Harvard, if the continents are to drift in the manner suggested by A. Wegener in Germany early in this century and supported by recent paleomagnetic evidence, then several hundred kilometers of the mantle must move with the continents. There is here some inconsistency with the evidence that under the continents, as well as under the oceans, the region of low velocity, the weakest layer, lies above 200 kilometers.

Models of the chemical and mineralogical composition of the Earth are usually based upon the composition of meteorites, for these bodies are thought either to have been formed at the same time as the Earth, or to be parts of a larger body formed at the same time as the Earth that broke up later and littered space with debris.

But the Earth itself gives us an opportunity to study the rocks of its interior. Diamonds are found in pipes in a matrix of mashed-up material which has clearly been forced through the crust almost explosively. Since we know that the diamonds must have been formed at depths of 100 or more kilometers below the surface, we can infer that the other lumps of rock found in the pipes also came from the mantle. We are on sure ground then in believing that the mantle rocks are a mixture of garnet periodotite and eclogite, but the proportions of each and the relation between them are still the subject of speculation. In fact, we have a large number of samples of the material below the Mohorovičić discontinuity already, long before man has been able to drill to this region and to secure samples from it.

We know well that the rocks of the crust have been milked from the upper mantle, and in fact we see this process going on today in all the volcanic regions of the Earth. In the same process the gases of the atmosphere and the water of the ocean evolved from the mantle. In this process, the silica which occurs in the mantle rocks as silicate (i.e., tied to some metallic ion), separates out and becomes the quartz of the granites of the crust. But we do not yet fully understand the process by which the separation occurs.

Laboratory experiments at high pressure and temperature based on the pioneering work of Percy Bridgman of Harvard have shown that at high pressure and temperature many materials suffer phase changes: when the pressure becomes high enough, they change to new forms of higher density and with different elastic properties. Because the temperatures at corresponding depths beneath the oceans and continents are so different, it is probable that these phase changes occur at different depths under the continents and oceans and thus account for the differences which the seismologist sees between the upper mantle under the continents and oceans.

What of the forces which have shaped the surface of the Earth as we see it today? There is no question that these are considerable. The energy required to produce uplifts of several kilometers over regions of thousands of square kilometers is enormous. A large earthquake releases seismic wave energy of the order of 10^{25} ergs.

A very large power station, of, say, 500 megawatts, produces this amount of energy in 70 years. Broadly speaking, there are three ideas with regard to the processes which produce these gigantic effects.

The first, and the most popular today, is that there are giant convection cells within the mantle. The convection theories face a serious difficulty over whether convection will proceed through the phase-transition boundaries which almost certainly exist in the upper mantle, and there are other difficulties. A second and older concept is that the forces which shape the Earth arise from the slow contraction of the Earth as a result of cooling. But there is now some question as to whether the Earth is expanding or contracting. A third possibility is that the energy for the Earth-shaping processes is derived in some way from the energy of the rotation of the Earth. But none of these hypotheses is free from objection, and there is no doubt that we have much to learn about the fundamental processes which have shaped the Earth.

We know a little about the interior of the other satellites and planets. From astronomical data we can determine the mass, and thus the mean density, of the planets. For the Moon, the mean density is so low that it cannot have a core like that of the Earth. For Mars the core, if present at all, is relatively small, and it has been conjectured that the absence of a magnetic field on these bodies is related to lack of cores in both bodies. The Mariner results show that the Martian atmosphere differs from that of Earth by having much less water, and much less oxygen, so it is conjectured that the differentiation processes in which the crust and atmosphere evolve from the mantle have been smaller in scale on Mars than on Earth. It is estimated that the crust of Mars is several hundred times thinner than that of Earth. In fact, in spite of the difficulties of space exploration, it may well be easier to drill a Mohole on Mars than on Earth.

Increased knowledge of the interiors of other planets will be gained in the decades ahead by the use of spacecraft. Such insights will be significant to us as we attempt to understand the nature and history of the solar system, and here comparisons between the various planetary interiors will permit us to make generalizations. Yet the Earth itself will continue to hold our keenest interest, not only because it is our abode but because our better understanding of it is crucial to our interpretation of data from other planets.

• 7

The Crust and Continents

EUGENE HERRIN

The development of society has been, and is, strongly controlled by the natural resources available to man in the upper part of the continental crust, from the soil which produces most of his food to the petroleum he takes from holes drilled many kilometers into the Earth. The importance of natural resources to man's development is reflected in the names given to the ages of civilization: the Stone Age, the Age of Copper, the Bronze Age, and the Iron Age. No wonder, then, that man has shown great curiosity about the skin of the Earth upon which he lives.

The continental crust is generally 30 to 40 kilometers thick, extending from the surface to a rather sharp break, the Mohorovičić discontinuity, which marks the boundary between the crust and mantle of the earth. The upper part of the continental crust is in many places covered by layers of water-lain or wind-lain deposits called sedimentary rocks or by their metamorphic equivalents—that is, sediments greatly altered by the increased temperatures and pressures of deep burial. These rocks were later exposed by the action of erosive processes. Below the uppermost layers, the crust is composed of material having the properties of granite. With depth the crustal rocks become more dense, more rich in iron and calcium, and depleted in silicon. At the base of the crust these rocks have the properties of the dark lavas called basalts that are found in many parts of the world. The oceanic crust is much thinner and consists only of rocks of the basaltic type.

The crust and upper mantle do not have sufficient strength to support the great load of mountain blocks or upland plateaus over an extended period of time. These highlands would sink slowly into the underlying material until an equilibrium was established. The required flotation process can result from a depression of the lower boundary of the crust, the Mohorovičić discontinuity, into the denser mantle or it can result from the presence of material of anomalously low density in the upper mantle below the discontinuity. The latter process can be shown to be statically equivalent to a depressed and thickened crust. It now appears that both phenomena are important in maintaining the flotation of continental highlands; thus, we expect a thicker, deeper crust beneath the great mountain systems and, in fact, the thicknesses are as great as 75 kilometers in Central Asia. Under lowlands a more normal continental crust is found. Under the oceans, the crust is only a few kilometers thick.

Geologists and geophysicists whose job it is to explore and try to understand the properties of the crust have over the last century and a half developed a fascinating picture of its history and dynamics, a picture well described by Kirtley F. Mather of Harvard University:

"To get some idea of a geologist's view of the Earth and its history, an ideal instrument would be a time-lapse camera such as those used to photograph the progress of a flower from bud to full bloom. One can imagine a motion picture of the Earth, taken by a camera on a platform some thousands of miles out from the Earth's surface, with one picture of the same hemisphere taken every 5,000 years. After nearly a billion years, we would have a movie of truly epic proportions, telescoping a large part of the history of our planet into a 3½ hour drama.

"In such a motion picture the Earth would appear to be alive, its exterior writhing in spasms. Great wrinkles—mountain ranges and canyons—would appear in land that a few moments before had been flat. Lands previously covered by shallow water would emerge and other lands would be flooded by spreading seas. Glaciers and running water would file the mountain ranges into jagged peaks, then down to low hills and finally back to flat valleys once more. Green jungles would change suddenly into stark deserts. Great gashes, such as the Rift Valley in Africa, might open up in a few seconds. Volcanos would fairly jump out of the surface and then be worn away in a minute or two. Vast ice sheets would expand over immense areas then retreat and expand and retreat again, carving the land and leaving behind new rivers, lakes and soils. This is the kind of rapidly changing world the geologist sees."[1]

Scientists have not always viewed the Earth as a changing body. Prior to the beginning of the nineteenth century, geologists thought of the crust as rather static, the Earth having been created through a series of catastrophic occurrences and then having remained much the same for the rest of its life. In 1795, James Hutton of Edinburgh developed a theory which was to become central to all of geological thinking. Hutton said, "No extraordinary events are to be alleged to explain a common appearance. Chaos and confusion are not to be introduced into the order of nature, because certain things appear to our partial view as being in some disorder. Nor are we to proceed in feigning causes, when those seem insufficient which occur in our experience."[2]

In effect Hutton was saying that the small, seemingly insignificant changes which occur on the surface of the Earth, such as the wearing away of a stream bank or slight changes in a shore line, are in the vastness of geological time capable of completely altering the surface of the Earth and of producing the features we see today. Modern geologists follow Hutton's example by observing the natural processes that go on no matter how slowly and extrapolating back through time in order to piece together the history and development of the Earth's crust.

The processes of erosion, the wearing down of the land surface by the action of wind, water and ice, are visible to all of us. If these processes were to go on for extended lengths of time, the con-

---

[1]Kirtley F. Mather, *The Earth Beneath Us* (New York: Random House, 1964), p. 9.

[2]James Hutton, *Theory of the Earth* (Edinburgh, 1795), Vol. II, Part II, p. 547.

tinental highlands would be worn completely away, and through wave action the seas would eventually obliterate the continents altogether. Under these hypothetical conditions, which follow logically from Hutton's theory, there would be no land surface, no continents at all; man could not have evolved in his present form. But there are continents now and there have been continents throughout geological time. Clearly, there must be a process acting to lift up mountains and to preserve the continental masses in a condition of quasi-equilibrium with the destructive forces of erosion. In dynamic and sometimes catastrophic geological events (for example, earthquakes and the eruptions of volcanoes), we see evidence of such phenomena tending to maintain the elevation of our land surfaces. The study of the remarkable conflict of forces destroying and rejuvenating the highlands is perhaps the most fascinating part of the geological sciences.

Only in the last decade have we come to understand that the great topographic features of the continents, such as the mountain systems and the island arcs, are by no means superficial features but that their roots extend through the crust and into the upper mantle to a depth of several hundred kilometers. It is in this region below the crust that we must search for the origin of the forces that shape the crust itself and maintain the highlands against continual erosion. Some parts of the continental masses have floundered beneath the oceans and there are new segments of continents, but on the whole the main mass of the continental blocks has persisted through geological time although, as we shall see, they may have moved relative to the interior of the Earth.

The southern half of western California is broken by a long fracture, the San Andreas Fault, along which occurred the destructive San Francisco earthquake in 1906. The region close to the coast of California from San Francisco southward to San Diego is moving northward at a measurable rate. Geologists believe that the coastal section has moved several hundred kilometers during the last 100 million years. There are similar examples of fault motion in other continents. Following the principles outlined by Hutton, one can conceive of large segments of the continental crust moving relative to the underlying mantle; in fact, the continents themselves may have shifted great distances with respect to one another.

One may perform an interesting experiment in the following

way. Cut paper patterns of the European, African, and American continents to fit their representation on a medium-size globe. Now place these patterns on the globe and attempt to slide them toward one another. North America will move toward South America, thus filling the Gulf of Mexico and the Caribbean Sea. Britain and the Scandinavian states will fill the Baltic and North Seas, while the compression of Europe towards Africa will fill the Mediterranean. Moving the two sets of continents toward one another will fill the Atlantic. One observes an amazing fit between the eastern coastline of South America and the western coastline of Africa. Is it possible that we see here one large continent that has broken apart and drifted into the present positions?

There are other lines of evidence to indicate that continental drift of this type may have occurred. When lavas, extruded during volcanic eruptions, cool below a certain temperature, there is locked into their mineral structure a magnetic orientation which coincides with the direction of the Earth's magnetic field at the time of cooling. This magnetic memory remains fixed for the life of the rock although the magnetic field of the Earth may change from time to time. By analysis of the radioactive materials present in the rock, the date of their cooling may be determined. Sensitive instruments can be used to measure the direction of the residual magnetic properties of the rock which constitute, in effect, a magnetic memory of the Earth's field at the time the rock cooled. As these paleomagnetic data are collected from many parts of the Earth, it becomes possible to chart by triangulation the position of the Earth's magnetic pole as a function of time through the geological ages. The "fixes" so obtained from the different continents do not coincide, indicating that the present position of the continents is not the same as it was in the past. These studies strongly suggest that the continents have drifted with respect to one another during the last few hundred million years.

Yet continental drift of such magnitude presents difficulties. As we have seen, the major continental features are continued in the mantle, perhaps to depths of several hundred kilometers. If the continents have drifted, do the underlying mantle rocks drift with them? What physical forces could cause such large blocks of Earth material to move across the surface of the globe? Perhaps in all of the geological sciences there is no more intriguing idea than this of continental drift.

If the Euro-African and American continental blocks have drifted apart in the last few geological periods, then the Atlantic Ocean basin must be a rather young feature and should contain only relatively young sediments. It has recently been proposed that a comprehensive survey along a line from the east coast of the United States to the west coast of Mauritania in Africa be undertaken with sufficient drilling to determine the age of the sediments in the Atlantic basin. This experiment should provide critical evidence regarding the validity of the drift theory.

As I have already indicated, the crúst is separated from the upper mantle of the Earth by a rather profound discontinuity, a fairly sharp break at which the properties of the rocks change quite significantly. The density of the material increases, the seismic velocity increases, and probably the radioactive content decreases markedly at this boundary. The temperatures and pressures present in the crust and upper mantle of the Earth can be reproduced in the laboratory. We believe that the rocks present there are all represented by samples found at the surface and that by subjecting selected rocks to the required temperatures and pressures we should be able to duplicate the materials found in the lower crust and upper mantle. It is by this means that geologists and geophysicists have attempted to describe the composition and physical properties of this part of the Earth.

Several years ago the United States embarked upon a fascinating and ambitious experiment: an attempt to drill through the crust and the Mohorovičić discontinuity into the upper mantle and to bring back specimens of the material there. This experiment, called Project Mohole, has been deferred for the time being, but good progress was made in the design phase and sooner or later man will pursue it. Our technology is not sufficiently developed to make possible the drilling of a hole through the continental crust, which is 30 or more kilometers thick, but the oceanic crust is much thinner. A hole 6 or 7 kilometers deep can be expected to penetrate the mantle. The successful completion of Project Mohole should allow us to check our laboratory inferences concerning the composition of the crust and upper mantle and the properties of the unique discontinuity that separates them. Moreover, the engineering skills learned in this way would help us to explore more deeply into the crust for the raw materials required in ever increasing amounts.

Man is curious about the history of the crust on which he lives and also about the forces that shaped it. What causes the earthquakes and volcanic eruptions that in his lifetime give evidence of the unrest in the Earth's mantle? What forces built the mountains and preserved the continents against the attacks of erosion?

Measurable amounts of heat are known to flow out of the upper crust, something like one millionth of a calorie per second per square centimeter over the entire surface. In the nineteenth century it was thought that this heat flow resulted from the cooling of the Earth's interior. Clearly, such a process over the ages would result in a reduced temperature in the outer mantle, and hence the Earth would shrink, with the crust wrinkling like the skin of an apple drying in the sun. This contraction theory was called on to explain the tectonic forces which cause earthquakes and give rise to the folded mountain belts across our continents. With the discovery of natural radioactivity, however, it was found that most of the heat escaping from the Earth could be attributed to radioactive decay, and it was not at all clear that the Earth was shrinking: in fact, it might be expanding, a situation which would arise if more heat were liberated from radioactive materials than could be conducted to the surface and there lost by radiation.

Another theory very popular during the last decade suggested that the mantle of the Earth, when subjected to stresses over a long period of time, might flow in a ductile manner. Because the temperatures are higher in the lower mantle than near the crust, convection cells might develop, with the hot, less dense material rising to the surface and the cooler material sinking. The stresses imposed upon the crust by these motions could result in mountain building and continental uplift. Arguments regarding the convection theory are still going on. Many of these are very complicated and depend upon a rigorous application of thermodynamics and the theories of solid-state physics.

It is clear, however, that we do not yet fully understand the origin of the forces that have shaped our continental crust. Convection may play a part, but the picture is far from complete. Of one thing we can be fairly certain: By some means, by some trigger mechanism, the heat energy stored within the mantle is converted into stresses which bend or break the overlying crust and allow the continents over the geological ages to keep their heads above water.

Along with the broad problems concerning the history of the Earth and the causes of crustal deformation, there are many practical problems with which the earth scientist is concerned that are of utmost importance to man and his society. In 1964 Japan and the United States entered into a cooperative program of study concerned with the cause and effect of earthquakes. It is to be hoped that such programs, diligently pursued, will someday lead to a technique for the prediction of earthquakes much like those warning systems concerned with hurricanes, typhoons, and tsunamis. Many areas highly subject to earthquake activity have large populations; disastrous earthquakes which have occurred in Tokyo, San Francisco, and recently in Tashkent illustrate the danger. Adequate warnings of impending earthquakes could save many lives and provide the opportunity for the protection of property. As we learn more about the crust of the Earth, we should be able to live more safely upon it.

Today the Soviet Union and the United States are engaged in ambitious programs for the exploration of the Moon and nearer planets. For many years this exploration will consist of brief visits by both manned and unmanned vehicles, with the collection of data greatly restricted by the inhospitable conditions of space. Sample points will be scarce, but we will wish to build a picture of the properties and history of these bodies from the information that will become available. Clearly, we must extrapolate from studies of the Earth's continental crust so that we can have guides for the interpretation of data from the planets. In exploring the crust of the Moon we must keep in mind similar regions on our continents, always making proper allowance for the lack of a hydrosphere and atmosphere on the Moon as well as for the extreme variations in temperature. A careful comparison of the properties of the lunar crust with the Earth's continental crust should be man's first scientific goal in the exploration of the Moon.

As we learn more about the planets, we should in turn be able to use this new information in our investigation of the Earth's geological history. Without water or an atmosphere, the Moon is not subject to the relentless processes of erosion found on Earth; nor shall we find there thick sedimentary layers obscuring the parent crust. The volcanic processes that were so important in building the Earth's continental crust should be much more clearly revealed on the Moon. The maria of the Moon as well as its highlands are available for study by geologists, for there are no seas as on Earth to cover a sizable fraction of its surface. Thus, our exploration of the planets and of the continents upon which we live are intimately related and must, if they are to be successful, complement one another.

The interrelation of studies seen on a solar-system scale is also true in the exploration of the Earth itself. We cannot hope to understand the reasons for crustal movement until we understand the properties and forces of the mantle well below the crust. The development of the continents is closely related to the development of the atmosphere and of the oceans, for without the interaction of the hydrosphere and atmosphere with the lithosphere of the earth, most of the rocks which form the surface of the continental crust could not have formed. It is thus apparent that the geologist must study the Earth as a planet, using all the tools and applied sciences available to him if he is to understand continental development. In the words of J. H. F. Umbgrove, late professor of geology at Delft in Holland, "Only few realize that the realm of earth science extends from the infinitely remote ages and depths of the universe to the origin and meaning of all organisms including the inmost depths of ourselves. Studying problems of earth science and examining their relation to other phenomena, the route inevitably leads to these two extremes of human thought. It does not matter where we start, because all phenomena appear to be interrelated and each portion of the universe will come up at a certain moment to play its own part. We may start with any given landscape, in our country or in the East Indies, in the Pacific or in Canada, in Italy or in Russia; we shall always end at these two extreme poles of thought."[3]

---

[3]J. H. F. Umbgrove, *Symphony of the Earth* (The Hague: Nijhoff, 1950), p. 1.

8

The Earth's Magnetism

WALTER M. ELSASSER

The Earth's magnetism is one of the oldest of nature's mysteries to arouse the curiosity of man. In spite of their age, the problems and questions relating to the Earth's magnetic nature have only in our own time proved accessible to a more detailed understanding. The magnetic compass was already used as a navigator's aid in the Age of Exploration. Without it, the achievements of a Columbus or Magellan are hardly conceivable. In the year 1600 the British physician William Gilbert published a book entitled *De Magnete*. He first enunciated the proposition that the whole Earth is one big magnet, that magnetism of the Earth is a phenomenon that pertains to the planet as a whole. It does not just apply to some particular geographical region.

The survey of the Earth's magnetism in this chapter falls naturally into two parts. First, I shall describe the factual knowledge of the Earth's magnetic field that multitudes of observers have slowly acquired. They have provided us with ever rising amounts of data and also with gradually increasing precision. This brief account will help us to understand the following section on the origin and physical nature of the Earth's magnetism.

The term "magnetic field" indicates the presence of a magnetic force at each point in space. This force can be measured with respect to both magnitude and direction. The compass employed to measure this force was in the beginning made by tying a piece of magnet to a piece of wood and letting the whole float on water. At present a compass is designed so that the needle can rotate in a horizontal plane by means of a delicate bearing. In a similar fashion it is possible to let the needle swing in a vertical plane. In this experiment one finds that the needle comes to equilibrium at a certain angle with the vertical. This angle is known as the "angle of inclination."

If the Earth were a uniformly magnetized sphere, such as one can make in the laboratory from a ball of steel, its magnetic field would be extremely simple. The magnetic compass would point to the north everywhere at the Earth's surface. Again, for such an ideal magnet, the angle of inclination would depend on the geographical latitude and would be independent of the longitude. It would vary from a strictly horizontal direction pointing due north, at the equator, to an exactly vertical direction at the poles. But such ideal simplicities do not reckon with the interesting phenomenon of magnetic anomalies or irregularities. These were first discovered nearly five hundred years ago. The early navigators, by looking at the pole star, found that the magnetic needle does not point exactly to the north but deviates from this direction, more or less, at different places on the Earth. This difference between the true north and the direction of the needle in the magnetic compass is called the "declination" of the needle. Depending on geographical location, declination can be either to the east or to the west. It may be as large as 25 degrees or more, although often it is much less. But the declination of the needle is not the only irregularity of the Earth's field. The inclination differs almost everywhere from the value that one can calculate for an ideally magnetized sphere. The difference is again more or less irregular and in its magnitude and general character quite comparable to that of the magnetic declination.

Looking at the Earth's field as a whole, it can be approximately described by a magnetized sphere, but only if the magnetic axis is inclined relative to the geographical axis by about 11 degrees. This model accounts for a major fraction of the difference between the true and the idealized field but not for all of this difference. What remains unexplained in this way is a field whose irregularities are of two types. One of them is highly localized, extending over regions only a few kilometers in diameter. This type of irregularity can be traced to local magnetized bodies of iron ore or to similar deposits. It is of little interest for us. The other kind of irregularity is always on a very large scale. These features extend over many hundreds, more usually thousands of kilometers, and such irregularities exist with respect to the inclination, the declination, and the absolute strength of the magnetic field. We see, therefore, that the Earth's magnetic field is far from being a simple phenomenon. The complexity of the observed facts will later appear as a result of the complexity in the physical processes which cause the field.

So far I have spoken of the variability of the Earth's magnetic field in space, meaning changes that we find as we travel along the Earth's surface. There is also a quite similar variability of the field in time. Such a change must be described by means of the length of time required for it—that is, we need to know something about the rate of the variation in time. There are certain minute changes that take place within a few hours or a day, but the most significant changes are large and also slow; they require times ranging from some tens of years to some hundreds of years. Since we have had very precise and reliable magnetic observations for well over a century and have some observations that go back several centuries, changes of this type can be clearly analyzed.

As we study the records, these changes appear related to the variation of the field in space. It soon becomes clear that we are dealing with one and the same phenomenon, the variability of the field, both in space and in time. Deviations from the idealized field grow, reach a maximum in the course of, let us say, a hundred years, and then decrease again. This occurs over areas perhaps the size of a continent, a few thousand kilometers across. It seems that such magnetic play has been going on for a very long time.

The angle between the Earth's magnetic axis and its geographical axis is also not constant, but it changes more slowly. This change involves times of about one to two thousand years. Observations on the magnetization of marine deposits show that this variation is also irregular. On the average, over some ten thousand years, say, this deviation from the geographical axis is random. The conclusion is that there is only an impermanent and transitory—not a systematic—difference between the two axes. A further remarkable fact is that the average magnitude of the field also changes with time. Reliable observations of the average magnitude of the field began around 1830–40. From then until about 1940 the total magnetization of the earth has decreased by about 3 to 5 percent, but since 1940 the mean magnetization has remained roughly constant. We feel quite confident that this surprisingly large change is real and not just a deception caused by inadequate data.

Looking now at this variability of the Earth's magnetic field, one would very much like to know what happened to the field at earlier times in the Earth's history. Geologists do not usually think in terms of thousands of years but of many millions of years. Scientific developments since the end of World War II have produced a remarkable tool for acquiring knowledge here. This tool is the study of the magnetism of rocks, also known under the name of "paleomagnetism." The basic idea is as follows.

Rocks are newly formed in two main ways, either by cooling of fluid volcanic magma or else by deposition of sediments in water. During these processes the rocks acquire a certain magnetization which is in the direction of the Earth's magnetic field as it existed at the time and place of rock formation. This magnetization is permanently locked into the rock. However, it is not the entire magnetization observed in actual rocks. There is a second, less stable magnetization produced by the field that exists now or has existed in fairly recent times. The success of rock magnetic studies dated from the moment when it became feasible to separate these two components by a suitable treatment of the rock, using a combination of heat and magnetic fields. What remains after this treatment is the stable component. It has preserved the direction of the Earth's magnetic field at the time of the rock's formation. Paleomagnetism is fast becoming a most useful tool of the geologist because it allows us to study slow displacements of the Earth's crust over geological ages.

Equally dramatic and of very direct concern here is another story told by paleomagnetism. At some moments in geological history the magnetic field has suddenly reversed itself. What was previously the magnetic north pole suddenly became the magnetic south pole, and vice versa. This can be clearly observed in some regions of the Earth where there has been a series of lava flows spread intermittently over a geological period of some length. Each lava flow forms a sheet of rock on top of the previous ones. A classical case of this is found in Iceland. Some of these flows show a permanent magnetization roughly in the direction of the present Earth's field. But there are others, sandwiched in between the former ones, that show exactly the opposite magnetization. Many examples of such field reversals are now known from rocks that have come from all parts of the earth. It seems that such sudden field reversals have taken place throughout the known geological history. They occur at quite irregular intervals but on the average are some millions of years apart from each other. The most recent such reversal has been closely studied; it took place 700,000 years ago. Unfortunately, we know little else about these strange reversals beyond the fact of their existence.

I come now to the main topic—namely, the physical origin of the Earth's magnetism. Since the observational facts have been known for so long, many hypotheses have been advanced and abandoned in the course of history. One such unacceptable idea is that the magnetism of the Earth is due to permanent magnetization of some materials, after the manner of a steel magnet. This idea is contradicted by the tremendous variability of the field, including reversals. For this and for a variety of other reasons, the idea of static magnetization (or of ferromagnetism) is now completely abandoned. The second attractive but false hypothesis is that the laws of electromagnetism as we study them in the laboratory might not apply in the large dimensions of the Earth. For a long time this seemed a natural hypothesis to make and there has been a great deal of detailed research on it over the years. The final result, however, is quite definite: All the data indicate that the laws of electromagnetism in large dimensions are exactly the same as they are in small ones, and no basically new principles are required. What we do need is an adaptation of known principles to the large scale.

If we exclude static magnetization, there is only one known way to produce magnetic fields. Such fields always accompany electric currents. Any electric current is surrounded by a magnetic field. This fact has long been known and is described in much detail in every textbook of physics. We see, therefore, that the question of the sources of the Earth's magnetic field can now be posed in an alternate form: Where can we find the required electric currents, and what produces them?

For many years the explanation of the Earth's field has been hampered by the fact that it seemed such a unique phenomenon. There was a tremendous gap between magnetism in the laboratory and magnetism of the entire Earth. All this changed radically when the astronomer George Ellery Hale, founder of the Mount Wilson Observatory, discovered early in this century that all sunspots contain magnetic fields. On the average, a sunspot field is nearly a thousand times stronger than the Earth's field. Furthermore, many sunspots have a diameter hundreds of times the diameter of the Earth. These figures should give an idea of the colossal energies involved here. Ever since the days of Hale, astronomers have explored the universe for evidence of magnetic fields. We know at present that there are magnetic fields everywhere on the Sun, of a quite irregular kind, but those outside sunspots are much smaller than those inside. We know that the planet Jupiter has magnetic fields. We know also that certain classes of stars possess magnetic fields as strong as those in sunspots, and these often cover a very large part of a star's surface. We also know that the rarefied interstellar gases found almost everywhere in the universe carry magnetic fields. Hence magnetism is a universal phenomenon of cosmic physics. In all the cases mentioned, scientists have never been seriously in doubt that the magnetic fields are the result of electric currents that flow in cosmic matter.

These developments have lifted the Earth's magnetism out of its isolated position. It now appears intermediate in size between the laboratory and the stars, but in character it is very close to the magnetism of general cosmic matter. We thus have in the Earth's magnetism a valuable sample, close to home, of these pervasive electric currents in the universe. We know that electric currents can flow only in a material which is a reasonably good conductor of electricity. There are two types of good electric conductors: first, metals, and second, ionized gases. Consider now the ionized gases.

Most gases in the universe are so hot that they are partly ionized and hence can carry electric currents. This applies to the matter in stars as well as to interstellar matter. The capability for carrying current—electrical conductivity—depends only on the degree of ionization. It does not depend on the density. Hence, even very rarefied gases can carry large currents, and this agrees well with the astrophysical data.

There is ionized gas on the Earth itself, namely, in the ionosphere. This region of the upper atmosphere extends from a height of about 50 kilometers outward for some thousands of kilometers. We know that electric currents are generated in the ionosphere in a complicated but at present rather well understood manner. The mechanism is driven by the Sun, partly through ultraviolet light and partly by clouds of very thin gas emitted by the Sun and reaching the Earth. Such upper-atmospheric currents make a contribution to the Earth's magnetic field, but only a small one, a few percent of the whole at most. The chief characteristic of this external field is that it is rapidly variable. Most of it will change with periods of 24 hours. There are also so-called magnetic storms which usually last a few days and which can be traced to violent eruptions at the solar surface. These phenomena are quite different from the very slow variations over centuries that characterize the main part of the Earth's field. I shall now examine this slowly varying main field in detail.

As early as 1830, the mathematician Gauss showed that most of the magnetic field originates inside the Earth if the laws of electromagnetism hold for the Earth as a whole. This being at present certain, we have to look for a place inside the Earth where electric currents can flow. Not until seismologists had studied the Earth's interior extensively could this place be identified. Since about the end of the nineteenth century we have known that the Earth consists of two principal parts, called the mantle and the core. The mantle is the outer part; it is made up of rock which is in many respects similar to the rocks at the surface. Chemically speaking, these are silicates and oxides. Such substances are electrical insulators or nearly so. Hence the mantle must be a very poor conductor; electrical currents flowing in it will be exceedingly small. Analysis shows that we can disregard the mantle as a source region for the magnetic field.

The central part of the Earth—the core—is set off from the mantle by a very sharp boundary. The core has a diameter of about 7,000 kilometers, a little over half the diameter of the Earth. We know also at present a good deal about the physical and chemical properties of the core. We know with certainty from seismological and other geophysical data that it is liquid, not solid. In addition, we have a good idea of the core's chemical composition: it is composed mainly of molten iron. Naturally, it is not chemically pure iron; there are some impurities in it, but we think it is well over 90 percent purely metallic—that is, iron and some nickel. This view of the core's composition was for a long time the subject of controversy, but at present the basic knowledge is altogether well established. The temperature of the core is not too well known, but it is probably in the neighborhood of 4,000 degrees Celsius. From all this we can estimate how good a conductor of electricity the material of the core might be. This turns out to be quite adequate to carry the currents that produce the observed magnetic field. We thus come to a definite conclusion: The main part of the Earth's magnetic field owes its origin to electric currents flowing in the metallic core.

There remains then one hard question: How are these currents produced? The basic mechanism is quite similar to the mechanism that produces all the large-scale magnetic fields in the universe: they all arise out of mechanical motions of the conducting fluid. The fluid, as it moves, is capable of amplifying the electric currents and hence their magnetic fields. This process has many similarities to the way in which large electric currents are generated by rotating machinery in conventional power stations. Therefore the idea that electric currents and their magnetic fields are generated by moving fluids has become known as the *dynamo theory.*

At first sight, this might seem a somewhat strange notion. Could we then not amplify electric currents by stirring liquid metal, say mercury, in the laboratory? In principle this is surely possible, but it would be quite difficult because the velocities needed are larger than are practicable. The velocities which are needed for amplification depend on the size of the system. They are high for small systems, but the larger the dimensions of the fluid, the smaller these velocities can be. So engineers will continue to produce electricity by moving wires rather than by moving fluids, except in so-called plasma physics where velocities are very high indeed. But in the very large volume of the Earth, and even more so in stars, amplification of magnetic fields takes place readily, even with slow motions.

We have seen that the variation in time of the Earth's magnetic field is a direct result of irregular motions in the Earth's core. We may speak of a system of large-scale turbulent vortices or eddies. Maps showing the magnetic variations have a remarkable similarity to weather maps. In a sense they *are* weather maps of the core showing the large-scale circulation there. One can estimate fairly closely the typical velocity of the fluid in the core: it is of the order of one meter per hour. Again one must not think of these vortices as minor variations in an otherwise regular and uniform flow. They are a very large part of the total flow. This is quite similar to the Earth's atmosphere, where the flow consists principally of a set of ever shifting large-scale vortices.

On closer study, the motions in the core and their magnetic fields turn out to be far more complicated than one would at first suspect. It appears that the magnetic fields are twisted by the fluid motions into very complicated patterns. We have good reason to believe that inside the core the field is much larger than outside, perhaps as much as fifty times stronger than near the surface of the Earth. What we see at the outside is merely a sort of leakage of the internal and heavily twisted field which the complicated fluid motions generate and sustain by amplification.

We know from observations that the Earth's magnetic field is approximately lined up along the Earth's axis. This becomes intelligible if one takes into account the importance of the Earth's rotation for the character of the fluid motions. The rotation provides the controlling mechanical influence upon these motions and hence upon the field. The result is the creation of vortices of certain preferred types. As a result of this we find an average symmetry in the magnetic fields about the Earth's axis.

Another remarkable aspect of the amplifying mechanism in the core is its general instability. This expresses itself, for instance, in the field reversals. Although we understand nothing of the mechanics of these reversals, the very fact that they occur indicates clearly enough dynamic instability. Similar but more frequent reversals have been observed in the magnetic fields of some stars. This bespeaks again a rather widespread instability, in this case in the processes that generate the cosmic magnetic fields. Irregular reversals also occur in laboratory experiments with so-called disk dynamos. These are simple devices in which, in place of the moving fluid, there is a pair of rapidly rotating disks, electromagnetically coupled to each other.

Finally, there remains the question of the source of energy for motions in the core. Here we are not altogether sure. We know, however, that the magnetism of a fluid is quite different from that of a solid. In the fluid, the maintenance of motions and field requires only very small energies. We are almost certain that the motions in the core represent what is called thermal convection. As the Earth very slowly cools, heat is delivered from the core to the mantle. Since motion is by far the most efficient form of heat transport in a fluid, motion does take place in the core. The hypothesis of thermal convection as creating the motions seems very satisfactory but has not yet been established beyond all doubt.

Looking back now over these explanations, we recognize three main properties as basic for the Earth's magnetism. One is the electrical conductivity of the core; a second is fluid motion, specifically thermal convection; the third factor is the rapid rotation of the Earth. This last factor imposes symmetry upon the field and aligns it along the Earth's axis. If there were no rotation, there might still be magnetic fields, but even such modest regularity of the Earth's field as we observe would then disappear.

Because cosmic magnetic fields are generated by means of mechanical motions, we should think that the character and symmetry of such fields reflect those of the motions. This idea is borne out by observations. For instance, we know that there are magnetic fields in the gas of the spiral arms of our Galaxy. The fields are parallel to the direction of the arms, and this confirms our idea that the gas streams gradually outward in the arms. Next, let us look at magnetic fields of planets other than the Earth and of stars other than the Sun. Take the stars first.

Here, the pioneering observations of Horace Babcock at the Mount Wilson Observatory proved that stars with strong magnetic fields have two characteristics in common. They have a layer of intense convective motion; in addition they rotate very rapidly. Our Sun also has a convective layer, but it turns slowly, with a rotation period of 25 days. As a result the magnetic fields are irregular and scattered. If the Sun were far enough away to be just another star in the heavens, its overall magnetic field would be too weak to be observed. Many astrophysicists believe, partly on this basis, that magnetic fields in the universe are even more widespread than astronomical data so far indicate.

Now, as to some of the other planets. We know from radio observations that Jupiter has magnetic fields, but the average density of Jupiter is so low that we are certain it consists almost entirely of hydrogen. There should be only a small fraction of rocks and iron at the center. Deeper down, the hydrogen is strongly compressed and at the same time hot enough so that it is ionized and becomes a good conductor of electricity. Jupiter rotates very fast; its day lasts only 10 hours. Although we do not know the way in which magnetic fields on Jupiter are generated, the basic mechanism is probably similar to that in the Earth's core.

Rocket probes have come close to both Venus and Mars and have radioed back magnetic measurements. To our suprise neither of these planets has a magnetic field comparable to that of the Earth. If such fields exist, they are at least a hundred times smaller than the Earth's field. We can conclude that there is no motion in the cores of these planets or, much more likely, that they simply have no cores. This means that iron has not separated from the rocks as it has in the Earth at an early stage of its history. Mars rotates about as fast as the Earth but it is only one tenth as massive. This means a much lower force of gravity and could account for the fact that no core has been formed. Venus is nearly as large as the Earth. It is covered with a very heavy atmosphere, and this keeps us from observing its rotation. In any event, space exploration shows that there is great individual variation even among the inner planets. The study of cosmic magnetism will continue to play a role in the study of the solar system and in the efforts to understand its long and involved history.

9

The Ocean

W. S. VON ARX

Of the four minor planets of the solar system, Mercury is the hot planet, Venus the cloudy planet, Mars the arid planet, and Earth the watery planet. The Earth is unique in that, while it has clouds in its atmosphere like Venus and deserts like Mars, it shines with a blue light, as one can see when pale blue earth light fills the dark disk of the new Moon. The blueness of the Earth comes from two features: its atmosphere, for the color of the air is blue, and its ocean, for the color of clear sea water is also blue. I use the word "ocean" in the singular because the ocean basins are all interconnected and seawater is remarkably the same everywhere on Earth.

We can consider the world ocean to envelop the Earth as petals do a flower. There is the Pacific petal (the largest of all), the stubby Indian Ocean petal, and the long, narrow Atlantic-Arctic petal, all joined in a common center around the Antarctic. For all its wide expanse the world ocean is very thin, having about the same proportions of breadth and thickness as the paper map on a classroom globe.

The part of the ocean that is most familiar to us lies along the shorelines of the world. Here we see breakers smash against rocky cliffs or rush up the sands, and we realize that these waves could have been born in a storm thousands of miles away. Here too we can watch the tide rise and fall and realize that this deep breathing of the sea is caused by the far-reaching gravitational attraction of the Sun and Moon.

When oceanographers think of the ocean they characterize its depths in successive bands. The band beginning at the shoreline is the shallow water on the continental shelf, which is a seaward extension of the land itself and not very different from the land in its geological structure, composition, or mineral wealth. The continental shelves underlie about 7 percent of the total area of the ocean and support some of the major fisheries. In places they are even being mined for oil, sulphur, and minerals. This is probably the next frontier for industrial development as man intensifies his activities over the Earth.

Beyond the 100-fathom depth of the outer part of the continental shelf, the sea floor drops away to depths of 2,000 fathoms or more. This is the continental slope, where really deep water begins and where there is a structural transition from the thick continental crust to the thin crust of the ocean basins. A ship crossing this zone can pass from the shallow fishing banks into deep blue water in a matter of hours.

As we travel further seaward, the ocean grows deeper more gradually and the sea bottom grows smooth and almost perfectly flat, with only occasional sea mounts to interrupt its monotony. These flat areas are great basins filled through geological time with the fine debris of wind-blown dust and the skeletons of tiny planktonic organisms that live near the sea surface. This quiet order may, in places, be torn asunder by a rushing torrent of mud and sand shaken loose from the continental slope by an earthquake; but quiet is soon restored.

Still further on, near mid-ocean, the ship may cross a great mountain range, equal in height to the Himalayas. This mid-ocean range has been traced across the ocean floors for some 65,000 kilometers and is probably the longest range of mountains on this planet. Some geophysicists think the mid-ocean range is the scar joining the plates of continental rock that seem to be sliding slowly but inexorably over the Earth as the deeper materials of the Earth readjust to their internal heat. We see the tips of the highest peaks of the mid-ocean range as islands, particularly in the Indian and Atlantic ocean basins.

At the same time that we observe changes of water depth beneath the ship, we also feel a change of climate. As the color of the sea, which tends to be greenish near shore, gives way to the luminous cobalt blue of the open ocean, one can in winter have left a snowy coast and within a day or two be walking on deck in shirtsleeves. This change is related to the circulation of water in each of the ocean basins, which tends to collect warm surface water from the tropics and store it in great pools in the subtropics and middle latitudes. These pools of warm blue water float in mid-ocean as enormous lenses on the colder waters beneath. Their presence has a powerful influence on the climates of the world. It takes a great deal of time to heat and collect the water in them, and it would take an almost equally long time to cool them down. Their thermal inertia acts as a flywheel on climate by keeping us warmer in winter than we would otherwise be. The most famous of these lenses is the legendary Sargasso Sea.

By and large, the waters of the ocean are very cold. Three quarters of the ocean volume is colder than 10 degrees Celsius. Only the surface layer is at a comfortable room temperature. Of course, near the poles the surface layer too is very cold or even frozen. But it is remarkable that most of the world ocean is not frozen and nowhere on earth is the ocean even close to boiling.

Water substance is a liquid at atmospheric pressure only through a rather narrow range of temperatures. The mean temperature of the Earth is on the low side of the middle of that range. This makes the Earth physically unique and also biologically ideal, for most of the life on Earth depends to some extent on the presence of liquid water.

Water is not confined to the ocean. It saturates the ground beneath our feet, is ponded here and there in lakes which are really outcrops of ground water and, most important of all, rises as a vapor into the atmosphere.

Evaporation of water into the atmosphere takes a lot of energy. The distillation of an inch-thick layer of water into the air requires two solid days of high-noon tropical sunshine (if no energy is wasted on other processes). Because the Earth rotates and the Sun is not always directly overhead, the actual time over which that much energy can be accumulated on the Earth's surface is extended to somewhat more than a week. There is on the average almost an inch of liquid water suspended as a gas in the atmosphere. When this water vapor condenses into cloud and rain, the heat energy previously required to evaporate it is returned to the air, and this blanket protects creatures living at high latitudes on the Earth from losing heat too rapidly to the space beyond the atmosphere.

One can think of the tropics as a boiler and of the polar regions as a condenser in a great heat engine. Much of the water in the atmosphere is evaporated in the tropics, where sunlight is strong and relatively unseasonal. Some condenses in the tropical atmosphere and falls back to Earth, but some is moved by the winds to cooler climates where the heat of its condensation makes these cool areas less cold then they might otherwise be. It is the cyclic motion of water in the oceans and atmosphere which determines many of the characteristics of climate on the earth. Without its blue air and blue water, the Earth would be a very different place and possibly uninhabitable to the forms of life that are familiar to us.

Because of the very important links between the ocean and the atmosphere—the water cycle of evaporation and rainfall, and the currents of the ocean that are much affected, if not actually driven, by the winds—it is difficult to make very much sense out of one fluid without considering the influences of the other. The so-called air-sea interaction problem is one to which oceanographers and meterologists give a great deal of thoughtful attention. Contemporary studies of the circulations of the oceans and atmosphere are conducted in two ways: first, as a problem in physical geography in which the properties and motions of air and water in motion are simply mapped and described; and, second, as a problem in theoretical physics in which the equations of motion are employed in a variety of ways to produce simple theoretical models that hopefully bear some resemblance to natural processes.

In general, it is the whole picture that is being sought. During the past century or so man had to be contented with numerous small samples which he could assemble into a larger view. Today, because of improved means for traveling over and above the earth and for measuring its properties rapidly, the scientific question has changed from one of synthesis to analysis. The analytical problem is also supported by the growing facility we have gained in handling large-scale calculations with computers.

An example of this change can be found in the case of numerical weather prediction. Lewis Frey Richardson, the founder of numerical weather prediction, used the hours of waiting for casualty calls while an ambulance driver during World War I to predict the weather tendencies for one six-hour period over Western Europe and the British Isles. He estimated that it would require 64,000 people working continuously with pencil and paper just to keep abreast of the weather over the whole Earth by numerical methods. Now high-speed computing machines can outstrip the weather over the whole Earth by a factor of nearly 10 to 1.

But machines are not the whole answer. Machines have to be told what to do. We cannot predict the weather satisfactorily or anticipate the behavior of the ocean in response to wind and sunshine unless we have a physical understanding of how these things are linked. This quest for understanding is a challenge for true genius. A conception which synthesizes the bulk of human experience with the motions of the sea and air, allowing us to deal with them in simple general terms, may involve steps equivalent both to Newton's invention of the calculus and to his productive assertion that force equals mass times acceleration. These produced an intellectual explosion in classical physics. A great generalization of the same caliber in geophysical fluid mechanics could have equally far-reaching consequences and could be of immense importance to the welfare of all mankind on this planet.

But as Newton put it, he saw farther because he could stand on the shoulders of the men before him; we are now building the platform upon which another of his breed may stand. In constructing this platform we are using every technical device at our disposal. The space age has ushered in a number of orbiting sensors with which we can look at the atmosphere from the outside and measure the radiant energy it returns to space, see the distribution of clouds and the structures of storms, and even make some estimates of the winds at various levels of the atmosphere. These observations with scientific satellites provide us with a surveillance of the whole Earth, including the behavior of the atmosphere over the oceans, which heretofore has not been well known. The appearance of the atmosphere over the polar regions is also available to us now. Through the automatic picture-transmission system, much of this information is freely available to ground observers for consideration in daily forecasts; and the more subtle qualities of the atmosphere are available after analysis in published form, so that meteorology has a new dimension in the quality and extensiveness of the data available for its work.

Oceanography has not yet reached this state. From satellites one can see the surface of the ocean and look into its depths no more than a hundred meters or so with visible light. With infrared light one can look only into the upper few microns, and the range is equally short with radar wavelengths. With orbiting vehicles, then, it seems unlikely that oceanographers will have an opportunity to sense very much about the interior of the ocean, but it has been thought that much of what goes on in the interior of the ocean is expressed by changes in the surface elevation of the sea.

For example, it is well known that the sea surface rises and falls in a periodic way, owing to the gravitational attraction of the Sun and Moon. The tide is well known around the shores of oceans but is all but unknown in the great expanses far from continents and islands. It is also known that the sea surface yields to the pressure of the atmosphere upon it. This is called the "inverted barometer effect," for when the atmospheric pressure is high, the sea level is depressed. The sea yields by one centimeter for every millibar change in the pressure of the atmosphere upon it. When the wind blows across the sea surface, not only are waves formed but the sea tends to be piled up against barriers, and under violent conditions great broad waves called "storm surges" may be developed. Earthquakes near the sea can produce other waves of great length which sweep across the Earth at high speeds, very often as great as 400 knots, or very close to the speed of jet aircraft. These great waves, or tsunamis, are of immense importance to those who happen to live in their paths.

In addition to the waves that travel across the sea surface, there are bumps and dents in the sea level associated with the variation of the field of gravity over the Earth. Over the trenches which border the island arcs there tends to be less than the normal amount of gravitational pull; in these regions of deficient gravity the sea surface is depressed. Where there is an excess of gravitational attraction due to a continent or mountain range, the sea surface tends to be elevated above normal. A study of these elevations and depressions can reveal something about the internal structure of the Earth's crust beneath the oceans. Although the same information can be obtained by gravimeter measurements from surface ships or submarines, the gravimetric approach is very time consuming. It would therefore be helpful to sweep the Earth at satellite speeds if the same kind and quality of information could be obtained.

In still more subtle ways, the sea surface departs from level where there are ocean currents of a more or lesss permanent kind, such as the Gulf Stream, the Kuroshio, and the great Antarctic circulation. These changes in elevation arise from the fact that the principal ocean currents flow along the boundaries of very large masses of water that are conditioned by climate. For example, in the case of the Gulf Stream, the flow is from the equatorial Atlantic into the polar seas along a narrrowly defined route on the western side of the North Atlantic. On the right-hand side of the Gulf Stream one finds the Sargasso Sea, a tremendous expanse of rather salty and quite warm seawater. On the left-hand side of the Gulf Stream lies the Slope Water, which is moderately cold and relatively fresh. The Slope Water is like the water underneath the Sargasso Sea at a depth of about one kilometer. Therefore, it may be imagined that the water of the Sargasso Sea floats, as mentioned before, like a huge oil drop on the colder and fresher waters beneath. As with an oil drop, the surface of the Sargasso Sea should stand a little higher than that of the Slope Water on the left-hand side of the Gulf Stream. The difference in height has been computed to be about one meter, but no one has yet been able to measure it with any degree of accuracy. Were such measurements possible, however, it would then be far easier to determine the character of flow in the Gulf Stream.

It has been suggested that satellites might help to untangle this problem, as well as those of waves and tides and atmospheric pressure changes, by using an orbiting microwave altimeter to sweep the ocean surface and measure its height relative to the orbit of the satellite. While the orbit of a satellite is still too uncertain to be used as an origin of measurement, our technical competence in determining near-Earth orbits has grown enormously in recent years and promises still further improvement. Therefore, in a decade or two this may be a reasonable proposal for space technology to contribute in highly meaningful ways as a tool of oceanographic research. By going over and over the Earth one might hope to see the changes in the internal structure of the oceans, as they are revealed in its surface elevation, and by departures from the norm to assess the passage of traveling waves, like tsunamis and the slower-moving elevations and depressions of the sea surface occasioned by the pressure of the atmosphere, which in turn could provide useful information about the surface winds in and around storm centers.

For example, one of the most pressing questions in the minds of the meteorologists and oceanographers who deal with the air-sea interaction problem is that the sea and air interact most in centers of atmospheric storm. In a hurricane at sea one wonders where the sea ends and the air begins, for the air is full of water droplets and the sea is white with bubbles. Beyond the screaming of the wind one wonders what it is the ship finds to float in. This condition means that the area of the sea surface is enormously increased and that the droplets of seawater, driven by the winds for a time, return to the sea and thus carry some of the momentum of the wind into the surface layer of the ocean. Those droplets that evaporate release water vapor to the atmosphere and also small particles of salt which form the nuclei of rain perhaps at another place far distant. From a microwave altimeter study of the bump in the sea surface produced by the low pressure of the atmosphere in cyclonic centers, it is possible to compute the force of the winds and the pattern of circulation, which ultimately may be of vital importance to the prediction of tomorrow's weather by numerical methods. At the present time only the average oceans are represented in the numerical models, and this does not seem to be sufficient even in the light of our present understanding.

If we think, then, of the ocean and the atmosphere as a closely coupled system which is largely responsible for the distribution of heat and water on the Earth, then it is very much in man's interest to understand these systems. For example, if it were known where and how rapidly each storm at sea was moving toward the land, men at sea and ashore could avoid many unpleasant surprises. On a longer-range view, if the patterns of storm tracks could be known all around the Earth and the systematic departures from the normal be predicted in such a way that periods of drought or excessive rainfall could be anticipated, then man's life on this planet could be better arranged. Although this kind of physical understanding of our environment is far beyond our immediate grasp, the possibility of such understanding is now credible. Thus one is led to feel that, if we can at length come to grips with our own watery blue planet and see it whole through the use of orbiting sensors, we can look forward to the use of these same sensors in orbit around other planets, relying on the background of experience in interpretation of the Earth to draw meaningful conclusions about places where man has yet to set foot.

Thus far, I have considered some of the physical parts of the problems of oceanography and have neglected the equally important fields of marine biology, marine chemistry, marine geology and geophysics, and marine technology. It would take considerable space to discuss each of these, but it may help to show their relationships to physical oceanography if, in closing, I attempt to define oceanography itself.

Oceanography has been called the science of that part of the Earth covered by seawater. Oceanography may also be regarded as an assemblage of many ordinary land-based scientific disciplines whose purview has included marine research. But seawater is the central theme of the subject. Where seawater wets the solid crust of the Earth, oceanography enters the domain of geology. Where it reflects sunlight, is distilled into the atmosphere, or exerts a drag on the winds, oceanography is joined with meteorology. Where marine forms of life exist, or land forms migrate by way of the sea, oceanography merges with biology. And where man must combat or find uses for the sea or seawater itself, oceanography is allied with engineering and technology.

In all of these activities the chemist's role in the study of the oceans is nearly central, being as indispensable in the physical and biological pursuits as in marine chemistry. In a similar way biologists often work in close association with both chemists and physicists to determine the relationships between organisms and their environment. For all that the disciplinary training of each oceanographer may be dissimilar, a common bond is established by the ocean itself.

• 10

The Neutral Atmosphere

RICHARD GOODY

The region of the atmosphere lying below 50 kilometers contains 99.9 percent of its total mass, and all but a small fraction of the energy received from the Sun is absorbed there or scattered back into space. The central problem is to understand the complex interaction between absorbed solar radiation and a comparatively thin and light gaseous envelope. The absorbed energy causes some chemical changes, but for the most part only heats the atmosphere and gives rise to an intricate and variable pattern of motions. The motions, in turn, change the composition of the atmosphere, principally by changing the amount of cloud, and react upon the temperature of the ground and atmosphere. This circular process is completed, finally, by heat radiation from the Earth and its atmosphere out into space. On the long term, the cycle creates an overall balance of energy, so that our planet neither heats up nor cools down.

The term "neutral atmosphere" can be used to describe this region of the atmosphere. It is not a precise term, however, and only indicates that the problems connected with the destruction and interaction of atmospheric constituents caused by high-energy solar photons are treated in other chapters. The word "meteorology" may best describe the topic presently under discussion, although we shall explore ideas far outside the range generally considered under this heading.

Before examining some of the details of this gigantic heat engine, let us look at some general properties. The mass of the atmosphere is about one kilogram on each square centimeter of the Earth's surface, giving rise to a pressure or downward force of about 1,000 millibars. At a height of about 5.5 kilometers the amount of the atmosphere above is halved, and the pressure is also halved. Every rise of 5 to 6 kilometers leads to a further reduction by a factor of two. Thus the lowest 11 kilometers, which are of principal interest to the mid-latitude meteorologist, contain about 75 percent of the mass of the atmosphere; the lowest 20 kilometers contain 96 percent; the lowest 30 kilometers, 99 percent; and so on.

Most of the mass of the dry, neutral atmosphere consists of nitrogen (78 percent), oxygen (21 percent), and argon (1 percent). With one exception, discussed below, these gases are not split up by solar radiation below 50 kilometers, they are not ionized, and they do not partake in the heat balance either by absorbing solar radiation or by absorbing or emitting heat radiation.

The active constituents of the atmosphere are all relatively minor. Water vapor is the most important component below 10 to 15 kilometers. Not only does it give rise to clouds, but it releases large amounts of heat when it condenses, and its strong infrared bands absorb and emit heat radiation. Higher levels of the atmosphere are, however, comparatively dry. There the important constituents are carbon dioxide, forming about 0.03 percent of the atmosphere, and ozone, whose maximum contentration is about one thousandth of a percent. Both gases emit and absorb thermal radiation, while ozone also absorbs solar radiation in its strong ultraviolet bands.

I have mentioned the absorption of solar radiation, and the absorption and emission of thermal radiation. The absorption of solar radiation is a straightforward process. Outside the atmosphere a black surface, facing the Sun, absorbs about 2 calories per square centimeter per minute; that is enough heat to raise the temperature of a gram of water by 2 degrees Celsius. This amount of heat falls on each square centimeter in the tropics, where the Earth's surface faces the sun, but in polar regions the

Sun strikes at a glancing angle, and the heat is distributed over a much larger area.

This picture is complicated by the tilt of the Earth's rotation axis to its orbit, giving continuous day at one pole and continuous night at the other. Thus the overall picture is one of maximum heat input in the tropics, slightly less at the summer pole, and none at all near the winter pole.

Approximately 40 percent of this solar radiation is scattered or reflected back again into space by cloud, snow, sea, land surfaces and by the atmosphere itself; this amount has no influence at all on the heat balance of the planet.

A smaller amount (about 20 percent) of solar radiation is absorbed in the atmosphere itself and by clouds. Of this, a small quantity (about 1.5 percent) is absorbed by ozone in the region between 30 and 50 kilometers, where it has a particularly important effect because of the very low air density. The remaining 40 percent of the solar radiation is absorbed at the surface of the Earth, which we can consider to be the main source of heat for the atmospheric heat engine.

An engine also requires a heat sink, and this is the infrared heat radiation from the planet to space. Objects surrounding us at room temperature are continually exchanging heat radiation in large amounts. Because we normally gain as much heat as we lose, and because the situation is complicated by air motions, which also transfer heat, this heat radiation is not an obvious phenomenon. However, a body placed in space loses heat proportionally to the fourth power of its absolute temperature. For example, the total solar radiation absorbed by the Earth can be re-radiated by a black globe at about minus 28 degrees Celsius.

Why then is the Earth not at this low temperature? The heat radiation to space comes not only from the Earth's surface but also from clouds, water vapor, carbon dioxide, and ozone. Since the atmospheric temperature decreases with height, the average temperature of these emitting surfaces is, in fact, close to minus 28 degrees Celsius. The interposition of clouds and absorbing constituents between the Earth's surface and space shields the surface from the cosmic cold and returns to it a certain amount of thermal radiation. In general, therefore, the Earth's surface can be at a considerably higher temperature than minus 28 degrees Celsius.

Generally speaking, a cold gas above a hot surface starts to convect. Rising streams of hot gas mingle with downward streams of cold gas in such a way that heat is mixed away from the surface and into the gas. This process of convection, if sufficiently rapid, would carry all the excess of heat away from the Earth's surface into the atmosphere where it would be radiated away into space. However, the fall of pressure with height in the atmosphere introduces a new feature. As a bubble of air rises, it expands and cools by virtue of that expansion. Thus there is a certain rate of fall of temperature with height which will allow a bubble of air to rise without exchanging any heat with its surroundings. This gradient is known as the "adiabatic" gradient: it is a temperature fall of about 10 degrees Celsius per kilometer. If the atmosphere is violently stirred or allowed to convect, this is the gradient toward which the atmosphere will tend. Actually, the atmospheric temperature decreases by about 6.5 degrees Celsius per kilometer, the difference being due principally to the effect of water condensing in the form of clouds and rain.

The rather uniform temperature gradient I have described is a feature of the atmosphere up to about 11 kilometers in mid-latitudes, 17 kilometers over the tropics, and 0 to 8 kilometers over the poles, depending on the season. This lowest region of the atmosphere, which contains most of the important weather phenomena, is called the "troposphere."

At the top of the troposphere is the "tropopause," and here there is a striking and important change. The temperature gradient changes, with a jump, from a decrease with height to an increase over the tropics and to approximately constant temperature in mid-latitudes. The reason for this change is a great decrease in the convection, which is no longer able to control the situation and force an adiabatic gradient. Instead, the trend of temperature with height is mainly governed by an approximate balance of incoming and outgoing radiant energy in each small volume of the atmosphere. This state is called "radiative equilibrium."

The entire region from the tropopause to a height of 50 kilometers is known as the "stratosphere." The stratosphere differs in a number of important ways from the troposphere. As we have seen, mixing is less, a fact which makes itself felt, for example, in long residence times for radioactive debris in the stratosphere. Further, its composition is different in minor but important respects.

First, the stratosphere is very dry whereas the troposphere is nearly saturated: that is to say, the troposphere holds about all the water that it can without condensing, whereas the stratosphere holds only a tiny fraction of this maximum. Generally speaking, clouds do not appear in the stratosphere, and water vapor has almost none of the importance which it has in the troposphere.

Second, there are important photochemical reactions leading to the formation of ozone. Here we are considering the absorption of a very small amount of high-energy ultraviolet solar radiation by molecules of oxygen. A tiny proportion of the molecules is split up into oxygen atoms, which ultimately collide with oxygen molecules, forming new molecules consisting of three atoms of oxygen. These are ozone molecules. The number of ozone molecules is very small. Their maximum concentration is at 30 kilometers, where there is one ozone molecule for each hundred thousand molecules of air. But even these small concentrations are so significant that the region near 30 kilometers is sometimes called the *ozone layer*.

The importance of ozone lies in its ability to absorb about 1.5 percent of the solar radiation lying in the ultraviolet spectrum. (This has important effects on the Earth's life forms, which would not remain as they are if subjected to intense ultraviolet radiation; but that is something of a digression.) Most of this energy is absorbed high up in the tail of the ozone distribution, near 50 kilometers, where the atmospheric pressure and density are about 1 percent of ground values. The amount of energy involved is small, but it is put into a proportionately smaller mass of air than occurs near the Earth's surface. Because we have approximate radiative equilibrium, this heat can only be lost by increasing the thermal radiation, that is, by raising the temperature. Thus, in mid-latitudes the temperature rises above freezing point after falling to 70 or 80 degrees Celsius below freezing in the lower stratosphere.

This comparatively high temperature region is called the "thermopeak," and it exists at all latitudes and all seasons. The maximum temperatures are found in the summer hemisphere towards the poles, which is connected with the 24-hour illumination at these latitudes.

I now come to the topic which may be considered to be the central theme of meteorology: the motion of the air over the surface of the planet. It is a problem of unusual complexity, embracing a wide range of phenomena from small fluctuating eddies, a centimeter or two in size, to global wind systems. Between these two extremes we can first consider dust devils, convective clouds, thunderstorms, tornadoes, and sea breezes, all of which are insignificant on a global scale. Next, we have frontal disturbances, hurricanes, the traveling cyclones and anticyclones of mid-latitudes, the semipermanent high and low pressure systems over Iceland, the Aleutians, and the subtropics, and finally the huge Rossby waves.

Besides the sheer complexity of these motions, two simple but important concepts should be borne in mind. The first is that the basic drive must come from solar heating. Without the radiative sources and sinks which I have already discussed, there would be nothing to renew the motions as they are destroyed by friction with the Earth's surface. Thus the difference of solar heating between the equator and the poles drives the average global wind systems, which are known as the "general circulation" of the atmosphere. Differences of temperature between ocean and land masses cause the semipermanent highs and lows as well as the Asiatic monsoons. And on the smallest scale the sea breeze is driven by local temperature differences between land and sea.

The second important concept is the great significance of the Earth's rotation in modifying motions. The Earth rotates in a westerly (or eastward) direction, with a surface speed at the equator of 464 meters per second. Now picture a ring or torus of air surrounding the equator and an atmospheric wind system which attempts to move it towards the poles. If the torus is away from the frictional influence of the Earth's surface, it tends to conserve its *angular momentum* (the product of its velocity and its distance from the Earth's axis of rotation). As it goes toward the poles, this distance decreases to zero at the poles themselves, and the torus must accelerate to conserve its momentum.

Under the circumstances pictured, therefore, we may expect to find some very strong belts of westerly winds, particularly well away from the Earth's surface. Such westerlies are a prominent feature of the general circulation in mid-latitudes, although easterlies occur in both the tropics and in polar regions. There is also an important north-south component in the equatorial trade wind belts, with winds blowing towards the equator in both hemispheres. Because air cannot disappear, we conclude that it must rise in the convergence region.

As we go upwards through the troposphere, the most dramatic change is the extension and intensification of the mid-latitude westerly winds, which reach velocities up to 50 meters per second at the tropopause near latitude 30.

I have already pointed out that the drive for this general circulation must come from excess heating in the tropics, and we have seen that air rises over the tropics in the expected manner. Ultimately, in order to close the circulation and return air to the surface, the wind must spread towards the poles, generating strong westerlies. Unfortunately, although this simple picture starts to explain the observations in a promising way, we cannot explain all the observations by means of a single cellular circulation from equator to poles, carrying heat directly from source to sink.

It turns out that the motions in mid-latitudes are unstable and become erratic and turbulent. The turbulent eddies generated by the general circulation are the mid-latitude moving cyclones and anticyclones, which transfer angular momentum and heat from north to south by means of a vast mixing process. The coming together of cyclones and anticyclones in turn produces cold and warm fronts, with which are associated many of the detailed weather phenomena.

In the stratosphere there are also complicated wind phenomena of great interest, which are somewhat less well understood. The tropospheric winds have marked seasonal changes only on a local scale: the monsoon wind systems. The average winds generally blow in the same directions in winter and summer. Above 30 kilometers, however, the situation differs, and the prominent westerlies of mid-latitudes are present only in the winter; in the summer they are replaced by a broad belt of easterlies. Velocities can be very high, even ranging between 50 and 100 meters per second at heights near 60 kilometers. The reason for this wind system can be found in the heating of atmospheric ozone by the Sun which, we have seen, is strong near 50 kilometers. The heating is strongest near the summer pole and weakest near the winter pole, and there is a fairly steady decrease of temperature from summer to winter pole. In the troposphere, on the other hand, the tropics are hotter than the poles in summer and winter, which accounts for the characteristic differences in the seasonal winds.

Another phenomenon of interest has only recently been discovered. The textbooks used to state that at about 25 kilometers above the tropics the wind was either easterly (as measured by the drift of clouds from the explosion of the volcano Krakatoa) or westerly (as measured by von Berson). Averages taken over fixed months of the year showed no seasonal change at all. This apparent contradiction was resolved by the extensive data of the International Geophysical Year, which showed that the wind changes from easterly to westerly and back with a period of 26 months. With such a period, successive Januaries (for example) will show a variation from year to year, but averaged over many years the January wind will be the same as any other month. We do not yet have a complete explanation of this phenomenon.

Lastly, we have the spectacular seasonal changes over the polar regions. The stratospheric monsoon winds come lower in these regions. During the summer we have light easterly winds, rising to a maximum in the north in June. Westerly winds commence in September and rise steadily until February or March, when they can be as strong as 80 meters per second in a broad eccentric ring about the poles at a height of 35 kilometers. The spectacular feature of this circulation is the dramatic change to the summer regime, which takes place almost on a single day in February or March. The summer easterlies are re-established in a day or two, and during this time the temperature rises 30 to 60 degrees Celsius. This is the most cataclysmic change which has been recorded in the neutral atmosphere.

I must now return to a topic of vital importance to tropospheric meteorology, but which it is convenient to separate from its context: that is, the formation of clouds and precipitation. When air rises it expands and cools and, if it contains sufficient water vapor, the water condenses and forms clouds. The nature of the cloud depends upon the nature of the upcurrent: small-scale convection gives rise to the fair-weather cumulus whereas extensive sheets of stratus cloud accompany the slow rising in a depression.

The process is important for three reasons. First, there is the large amount of heat of condensation which is released, particularly in the tropics. Second, the varying reflectivity of clouds affects the Earth's radiation balance. And third, there are obvious social and economic interests in the occurrence of rain or snow at ground level.

The initial condensation takes place on minute motes of solid matter, known as "condensation nuclei." These number typically about 100 per cubic centimeter and, if all the available water is distributed among them, the average drop of water would not exceed a diameter of 50 microns (there are 1,000 microns to a millimeter, and therefore 20 drops of this diameter). Such drops, if they fall out of the bottom of a cloud, do so very slowly and evaporate almost instantaneously. They can form mist but not rain.

A raindrop that reaches the ground has a minimum diameter of 300 microns for light drizzle and a maximum diameter of 5,000 microns (5 millimeters) in a thundershower. It is interesting to consider why rain should form from cloud systems which are normally so long-lived. There are two important mechanisms.

The first requires that part of the cloud be at a temperature below the freezing point and yet still be liquid rather than ice. This supercooled condition is by no means uncommon; if a few particles of ice appear, there is a rapid distillation of water from the liquid drops to the ice, and a few large particles develop and fall to the ground, perhaps melting on the way.

The second mechanism requires a steady, relatively strong updraft, and an occasional large droplet of diameter about 100 microns, which may have grown by accidental jostling or by condensation on a very large nucleus. A large drop falls more rapidly than the small drops, which are swept past by the updraft and may impinge upon the large drop until it has grown to a diameter of 3 millimeters or even more. Such large drops break up into a number of smaller drops and the process starts again, thus concentrating the many cloud particles into a few, far larger raindrops.

The possibility of changing the behavior of clouds by artificial disturbances has attracted much attention in recent years. Local rainfall could perhaps be increased, hail damage diminished, and dangerous weather systems such as hurricanes modified. If the cloud is supercooled in part, such disturbances may be introduced by freezing a few drops either with dry ice or with artificial nuclei such as silver iodide. Some slight successes in this direction have been claimed.

The economic and social interest in meteorological phenomena resides almost entirely in the possibility of improved weather prediction, and it is unfortunate that this is by far the most difficult problem of the lower atmosphere. Thus, scientific advances in a difficult field are obscured from the general public by the difficulties encountered by local weathermen attempting to meet the public's demand for detailed weather prognosis.

The local television forecaster proceeds by empirical methods. Given some knowledge of the main atmospheric systems, he adds the detailed weather from experience. It is known, in general, what physical laws control the development and movement of cyclones and anticyclones, at least for periods of a few days. Detailed measurements covering the North American continent, interpolated and extrapolated with a judicious mixture of physical deduction and long experience, enable rather satisfactory prognostic charts to be prepared at different levels in the atmosphere for a day or two in advance.

A recent development of importance has been the simplification and statement of the essential physical and dynamic laws in such a form that the entire prediction process can be put in mathematical terms. Once this has been done, the preparation of a prognostic chart becomes a task for an electronic computer. The public has already benefited from this development and, by taking as much as possible of the process out of the area of subjective judgment and into the area of objective judgment, the prospects for incorporating improved physical understanding and better observational networks are much improved. Plans are now under way greatly to increase the observational networks in both hemispheres, largely based on remote soundings with balloons and satellites.

Finally, we may ask, if we learn to predict the weather accurately and consistently, will it also be possible to control it? The question is an open one. Successful cloud seeding is a type of weather modification which may be around the corner. Other local effects may be brought about by blackening large areas to produce enhanced convection currents, such as a glider pilot encounters and uses. More ambitious schemes involve partially blackening the polar caps and damming straits (e.g., Bering Strait) to influence ocean currents and sea surface temperatures.

This last suggestion, whether feasible or not, at least emphasizes the close relation between physical oceanography and meteorology. It is too frequently overlooked that the oceans and the atmosphere form a single dynamic system for redistributing solar radiation, and that each is strongly coupled to the other. Ultimately we shall have to study this joint atmospheric-oceanic system, but advances in this direction are a matter for the future.

What does the future hold for these studies? In such a complicated field, in which the main phenomena have been identified, and the probable physical causes recognized, a sudden breakthrough is unlikely. Improved observations and increased theoretical effort will lead with certainty to improved knowledge in all the areas which I have discussed. The main stumbling block at present is our lack of understanding of turbulent or fluctuating motions; but this is also holding back other areas of classical physics and engineering.

Unfortunately, the cost of better observations now runs very high because of the necessary closeness of stations and the frequency in time required for a meaningful advance in knowledge. The related problem of distributing, storing, and processing this information is also a formidable one. Cost accounting may soon be invoked to define the limits of complexity desirable for neutral-atmosphere studies.

One novel aspect of atmospheric studies may lead eventually to fundamental, new ideas. At least we shall hear much about it in the coming years. This is concerned with the atmospheres of other planets. Mars has an atmosphere similar to our own in some ways, but with a very low surface pressure and an unexpectedly high concentration of carbon dioxide. Venus is very different, with a surface temperature possibly as high as 300 degrees Celsius, and a complete, opaque cloud cover. Mercury has no atmosphere that we can detect. The outer planets Jupiter and Saturn are totally different from the inner planets, with very deep atmospheres composed of light gases, strange formations and color charges, tremendous winds, and perhaps internal energy sources, like a cool star.

Until now, atmospheric scientists have had but one system to study in detail. In the 1970's this number will change to perhaps four or five. Although problems will differ from planet to planet, the interplay of ideas can only be stimulating and provocative.

• 11

The Ionized Atmosphere

S. A. BOWHILL

The density of the Earth's atmosphere diminishes so rapidly as altitude increases that one might be tempted to dismiss it as unimportant above about 50 kilometers and to regard it simply as a vestigial tail to the lower atmosphere of the Earth. However, the upper atmosphere has a number of highly distinctive characteristics which set it apart both from the lower atmosphere and from interplanetary space. Below 50 kilometers, solar effects on the Earth's atmosphere are primarily thermal, exciting transport processes such as winds and turbulence and causing the highly complex phase changes we most commonly know as "weather."

Above 50 kilometers, however, to the thermal effects are added a greater complexity of chemical reactions. From the simple ingredients of oxygen, nitrogen, and a few trace elements, hundreds of compounds—some neutral and some ionized—are built, so that each height range has its own distinctive chemistry. Only in a few cases is this chemistry reasonably well understood, partly because of the limitations imposed by trying to interpret a relatively few observations made on the ground in terms of complex reactions at high altitudes. The recent advent of rockets and satellites as vehicles for carrying space experiments to high altitudes, which might have been expected to resolve many of the difficulties, in fact has revealed hitherto unsuspected phenomena and has not thus far led to a complete understanding of the upper atmosphere.

I have used the word "chemistry" in connection with the upper atmosphere because it is suggestive of the nature of the processes that go on in that region, which extends from some 50 kilometers out to several Earth radii away. Below 50 kilometers' altitude, ionization in the Earth's atmosphere may be neglected under most circumstances, and the weather we experience is a property only of the neutral atmosphere. Above 50 kilometers, no clouds are present as we know them in the lower atmosphere, and therefore the Sun shines equally brightly throughout the day. This is not to imply that the neutral atmosphere has no role to play of its own above this level; for example, atomic oxygen, generated from molecular oxygen by the dissociative action of the Sun during the day, recombines during the night, giving off as it does so a greenish glow visible from the ground, which we call the "airglow." As a matter of fact, this glow is the source of most of the night-time glow from the sky; what we think of as starlight is in fact principally this oxygen airglow. However, most of the spectacular effects occurring in the upper atmosphere are associated with charged particles—electrons and positive ions. The importance of this ionization leads us to use the term "ionosphere" in speaking of the ionized part of the atmosphere.

The word "ionosphere" implies that ionized particles are present in the upper atmosphere, and this is indeed the case. In fact, the electrons produced from ionization of the atmosphere by ultraviolet rays from the Sun make the upper atmosphere into an ionized plasma. This plasma has dramatic effects on electromagnetic waves of frequency a few megahertz and below. In addition, it is capable of passing a direct current and therefore can affect the magnetic field of the Earth. Indeed, the ionosphere around 100 kilometers in altitude is the region where the effects of movements of the lower ionosphere become entangled with movements of the magnetic fields in interplanetary space.

Here I should like to describe some of the developments of ionospheric investigation, some of the advances that have taken place in the past few years, and some of the puzzling problems which are currently before us.

Even in the nineteenth century, the existence of ionization in the upper atmosphere was suspected when it was found that the Earth's magnetic field appeared to change cyclically during the day, and it was found that the current system needed to explain these variations must be located outside the Earth's sphere—and therefore, possibly, in the atmosphere itself.

Curiously, however, the suggestion of an ionized layer was first made on incorrect grounds. Because it was known in the late nineteenth century that gases became electrically conducting at low pressures (that is, capable of sustaining an electric discharge), it was felt that the low pressures present in the upper reaches of the Earth's atmosphere would have a similar effect. At that time, of course, it was not known that the Sun gave off the very short wavelengths of radiation which are now known to comprise most of its spectrum and which, in fact, make the upper atmosphere electrically conducting.

A direct indication of the presence of free electrons in the upper atmosphere was given by Marconi's experiments in long-distance wireless telegraphy. Theoretical calculations of the field strength of radio signals propagating around the curve of the Earth had indicated that it should be impossible to communicate over distances of more than a few hundred miles, as the signals would fly off into space instead of following the curve of the Earth. However, when Marconi succeeded in transmitting signals across the Atlantic Ocean, it was evident that some effect was present which constrained the radio waves to maintain themselves in the vicinity of the Earth. The explanation was not far to seek, for it seemed reasonable to suppose that the lower surface of the region of free electrons might act as a mirror of radio waves in the same way as a metallic surface reflects light waves. The theory of this effect was soon worked out, and it led to the picture of a concentric, spherical, ionized layer surrounding the Earth, radio waves being trapped between the layer and the surface of the Earth.

If this reflective layer were capable of trapping radio waves in this way, it should also be capable of reflecting them vertically back from a transmitter at the ground. In the early 1920's G. Breit and M. A. Tuve set up what amounted to the first radar transmitter, radiating pulses of energy from the ground which were detected as reflections from the ionisphere after a lapse of time corresponding to the distance from the ground to the ionosphere and back. Almost at the same time, Appleton in England found ionospheric reflections in using a regular broadcast transmitter.

Intensive investigations of the ionosphere did not begin until its usefulness for radio propagation became apparent in the early 1930's. Frequencies above about 2 megahertz had been relegated to radio amateurs for experimental purposes because they were thought to have no commercial use. Long-distance wireless telegraphy was at that time carried out at much lower frequencies, where large and cumbersome radio transmitters were needed. However, radio amateurs succeeded in establishing contact with each other over very great distances, even with the quite low radio powers which were permitted them by government regulations. Beginning about 1930, therefore, high-frequency propagation by means of the ionosphere became the normal means of communication over long distances, both for broadcasting and telegraphy.

It was soon found that the frequency bands which were usable for this purpose varied from night to day, with season, and with the stage of the sunspot cycle. Some means of measuring and even of predicting this behavior was therefore vitally necessary. It was soon found that the upper limit to the frequency of propagation was then related to the peak electron density present in the so-called F layer of the ionosphere. It was also found possible to measure this electron density by means of a ground-based radar whose frequency could be varied; this device came to be known as an ionosonde. Networks of these ionosondes, capable of determining the properties of the ionosphere from the ground, were set up in various parts of the globe, and as many as two hundred stations have since been furnishing regular data on the ionosphere to central collection agencies which analyze them and issue radio propagation predictions. These predictions permit radio communication engineers to judge with considerable accuracy which short-wave band would be best for communication to any given place at any time.

Until the end of World War II, the ionosonde was about the only useful technique for probing the atmosphere. For several reasons, however, it had given only a very limited understanding of the physics of the formation of the layers. First, it was capable of detecting only one of the constituents of the upper atmosphere, namely, electrons; second, its sensitivity was extremely limited, capable of detecting only the narrow range of electron densities from about two hundred to one, whereas the range of electron densities encountered in the ionosphere exceeds a million to one. However, rockets and satellites have since made possible a whole range of additional experiments, including the sampling of the upper atmosphere by a sensor immersed directly in it and the use of radio techniques involving propagation from the rocket or satellite to the ground (or vice versa). Typical of the data which have been most informative are the identification of the positively charged particles, or ions, accomplished by measurements of their mass, the identification of the various minor constituents of the neutral atmosphere, the detection of upper-atmosphere wind motions by releasing luminous trails from rockets, the orbiting of a miniature version of an ionosonde in a satellite to sound the ionosphere from above instead of below, and so on. From these and many other measurements, a picture of the ionosphere has emerged which I will now briefly describe.

The upper atmospheric regions are identified by three letters of the alphabet: D, E, and F. Each region has its own distinctive mode of formation and has a different effect on the propagation of radio waves.

I have already referred to the F region, which is the layer responsible for the propagation of short-wave radio signals to great distances. This is the uppermost of the ionized layers, having its maximum height at about 300 kilometers above the Earth's surface. But it does not coincide with the region of most intense ionization production, a paradox which bothered many ionospheric theoreticians until its explanation was found. Most of the Sun's ionizing effect takes place at about 180 kilometers in altitude; above that, the ionization rate drops off fairly rapidly with altitude. However, the ionization density is always determined by a balance between the processes of production and loss of ionization: since the efficiency of loss of ionization drops off even more rapidly than the rate of production, the balance reached is one in which the electron density increases with altitude rather than decreases.

Of course, this is based on the idea that ionization is both produced and lost in about the same height region of the atmosphere. Such is the case up to a critical altitude of about 300 kilometers, when a quite different process—vertical diffusion—becomes important: the ionization tends to drift downward in the atmosphere and adds to the supply at lower altitudes. In this case, the result is a decrease in electron density with altitude above 300 kilometers, producing a peak in the electron density at that altitude.

The Earth's magnetic field has a very important effect on the ionospheric plasma. Electrons and ions cannot freely cross magnetic field lines, but instead gyrate around them at a rate called the cyclotron frequency (named after the high-energy machine which uses this effect). However, they are capable of moving freely along the lines of magnetic force, subject to the influence of gravity. This has two interesting consequences. First, on the magnetic equator where the lines of the Earth's magnetic field are horizontal, free vertical movement for the electrons is evidently impossible in the way described for the formation of the F layer. The equatorial F layer is therefore quite exceptional in many aspects of its behavior, and its theory is far from understood at this time. Second, because a magnetic field line which intersects the Earth's surface in the northern hemisphere will also intersect in the southern hemisphere (at a place called the conjugate point), it is possible for electrons with enough energy to stream from one hemisphere to the other, provided a source for them can be found. In fact, the electrons which are first produced by the action of the ultraviolet solar radiation have just enough energy to make their way from the southern to the northern hemisphere, a journey of many thousands of kilometers. These electrons have, as a matter of fact, been detected by their heating effect in the northern winter hemisphere when the Sun's rays first strike the southern end of the field line and the northern end is still in darkness.

The region of the ionosphere below 160 kilometers comprises the D and E regions. The E region, down to about 90 kilometers in altitude, was at one time supposed to be fairly simple in its behavior because its maximum electron density, at an altitude of about 120 kilometers, coincides with the maximum ionizing effect of longer wavelengths of ultraviolet radiation than those responsible for the F layer. The intensity of this layer was found to rise and fall exactly as the Sun rises and sets. However, this region is also the seat of a number of intermittent effects, none of which are properly explained at the present time. One of these, the aurora, is familiar in that it is visible from the ground at high latitudes as a very variable light, present as rays, sheets, or streamers, having almost every color of the optical spectrum. This light is energy emitted by atoms and molecules in the E region which have been ionized or excited by streams of electrons coming into the Earth's atmosphere. The ionization they form can be readily detected by using radio waves. At lower latitudes, the intermittent phenomenon known as sporadic E occurs.

This may appear as an additional layer of ionization, superposed on the normal E layer, usually very thin and in some cases having many times the E layer intensity.

Another puzzling feature of the E layer relates to the ions which compose it. From rocket measurements, these are known to include molecular oxygen and nitric oxide, ions which have also been studied in the laboratory. However, the rate of disappearance of this ionization in the atmosphere does not seem to agree with the rate of disappearance in the laboratory, a discrepancy for which no good explanation has been offered.

The lowest layer in the ionosphere, the D layer, is associated with the absorption of radio signals. Radio stations in the broadcast band, between 500 and 1,600 kilohertz, have a much greater range at night than during the day because signals received during the day are transmitted directly over the ground from the transmitter to the receiver, whereas those at night are reflected from the E layer and therefore "skip" to much greater distances. The reason why these ionospheric reflections, present at night, are absent during the day is the phenomenon of absorption. The nighttime E layer may be likened to a polished metallic mirror located concentric with the Earth; during the day, however, it ceases to be an effective reflector, becoming tarnished, so to speak. This is due to ionization in the D region, below the E layer, which is present during the day but absent during the night.

So little ionization is required to produce this absorbing effect that the precise density and distribution were unknown for many years and are still the subject of debate. Recently, however, radio absorption measurements made from rockets have given an accurate picture of its distribution. It appears that the altitude of about 80 kilometers is quite critical for the D-layer ionization. Below this altitude it completely disappears at night; on a few days in winter, a strange additional layer is seen which springs sharply out of the background D layer at just this altitude, the ionization density being greater by perhaps a factor of ten than on a normal day. This effect, called the "sporadic" D layer for comparison with sporadic E, seems to be associated with dynamic effects in the lower atmosphere. At least, a quite significant correlation has been found between the increase in temperature in the stratosphere and the presence of this additional layer in the D region. The exact nature of this relationship is being intensively investigated at the present time.

One of the most fascinating developments in recent studies of the ionosphere is the detection of trace constituents, which are present in a very small concentration relative to the entire atmosphere but which are sufficient to affect ion density very strongly. For example, the F layer is composed primarily of ions of atomic oxygen, produced by the photo-ionizing effect of ultraviolet solar radiation. However, above an altitude of about 1,000 kilometers, the ions of atomic oxygen give way to protons, which can also be thought of as ions of atomic hydrogen. These are present at all altitudes above this transition level, and this region of ionosphere has therefore been called the protonosphere. The protons are not thought to be produced directly by solar radiation, but rather by charge exchange between neutral hydrogen and ions of atomic oxygen. This reaction takes place below the protonosphere, which accordingly can be regarded as a large reservoir of ionization, into and out of which ionization flows at relatively slow rates. The detailed consequences of this flow are being intensively explored at the present time.

Other minor constituents are extremely important in the E region of the ionosphere, below 160 kilometers in altitude. Here, for example, it has recently been found that metallic ions such as magnesium and calcium constitute a specific type of sporadic E layer, associated with meteoric activity. In the D region, the lowest part of the ionosphere, it has been possible since about 1965 to identify some of the ions present. In addition to ions of nitric oxide, which were to be expected at that altitude, ions with a mass number of 37 have been seen, which have tentatively been identified as multiply hydrated protons of the type which are found when an acid electrolyte is dispersed in water. The origin of these hydrated protons (if indeed they are not an artifact of the rocket which made the measurement) is at present uncertain.

The science of aeronomy, which treats all of the phenomena which I have been describing in the upper atmosphere, is one of the new branches of science and cuts across many of the older classical areas. Physics, chemistry, mathematics, and engineering each play some part in these studies, and an interdisciplinary approach is the only one to follow to reach an understanding of the upper atmosphere. Because the same atmosphere covers the entire globe, it is also an area where fruitful international cooperation may be effected. Ionospheric physics was one of the primary disciplines both of the International Geophysical Year (1957-58) and of the International Years of the Quiet Sun (1964–65). The former took place at the maximum of the sunspot cycle, the latter at the succeeding minimum. With the approach of the next sunspot maximum it is anticipated that increased attention will be paid to the short-term effects, such as those of solar flares which are of such short duration that they could not be studied by the techniques available during the International Geophysical Year.

In recent years we have learned that planets other than our own have ionized atmospheres. Earlier in this chapter I referred to Earth-orbiting satellites that have been used to study our own ionosphere from above. These studies have considerably augmented our knowledge of the outer reaches of our ionsphere. But spacecraft have also told us a good deal about other planetary ionospheres; for instance, the Mariner probe flight past the planet Mars made it possible for us to measure, with some confidence, the electron density of the Martian atmosphere and its variation with altitude. It bore such a remarkable resemblance to the F layer of the terrestrial ionosphere that it has been possible to carry over the entire theoretical basis for the Earth's ionosphere to interpret that of Mars. It seems likely that the principal difference between the mode of formation of the Martian ionosphere and that of the Earth is in the molecular gas responsible for the recombination of the ionization: on the Earth, it is probably molecular nitrogen, while on Mars it is probably carbon dioxide.

Ionspheres are likely to be a general property of all planets of our solar system that have even a trace of atmosphere. Studies of our own ionosphere will certainly continue to illuminate our studies of other planets, and it seems possible that planetary studies will lead us to look for effects in our own ionosphere which may hitherto have gone unnoticed.

• 12

The Magnetosphere

W. I. AXFORD

The magnetosphere of a planet is the region in which its magnetic field plays a dominant role in controlling its environment. In the case of the Earth, the magnetosphere extends outward, beyond an altitude of 100 kilometers, to a distant boundary which approaches no closer than about 50,000 kilometers. It is believed that the remaining terrestrial planets (Mecury, Venus and Mars) and the Moon have negligible magnetospheres because their magnetic fields are relatively weak. Jupiter, on the other hand, almost certainly has a large magnetosphere, according to deductions based on the properties of its radio emission. No results of an equivalent nature are available for Saturn at present, although it seems reasonable to expect that it, too, has a significant magnetosphere. Nothing is known about the magnetospheres of the other planets (Uranus, Neptune, and Pluto).

The inner boundary of the Earth's magnetosphere coincides roughly with the E region of the ionosphere. Its position is determined by comparing the pressure exerted by the geomagnetic field with that of the atmosphere. At ground level, of course, the magnetic pressure is relatively small, being only a few billionths of atmospheric pressure; however, the latter drops off very quickly with increasing altitudes whereas the magnetic field decreases quite slowly. Strictly, the two pressures become equal at an altitude of 130 kilometers, but as the effects of the magnetic field begin to be significant at 100 kilometers, the magnetosphere is usually considered to begin there.

The outer boundary of the magnetosphere, called the "magnetopause," has a shape which is roughly hemispherical on the sunward side of the Earth, and extends in the form of a long cylindrical tail to very great distances directly away from the Sun. In some respects then, the magnetosphere looks very much like a comet, with the Earth as its nucleus. Indeed comet tails and the tail of the magnetosphere are directed away from the Sun for the same reason—namely, because they are blown that way by the solar wind.

At the Earth's orbital distance from the Sun, the solar wind is a highly supersonic stream of ionized gas or plasma (mostly protons and electrons) which flows radially outward from the Sun at all times. The density of the solar wind is very low, being of the order of 10 particles per cubic centimeter. However the velocity is large—typically about 500 kilometers per second—so that the wind exerts quite a substantial ram pressure on the whole sunward face of the magnetosphere. By equating the ram pressure of the solar wind to the pressure of the geomagnetic field at the magnetopause, it is a simple matter to calculate the minimum distance to the magnetopause on the sunward side. This distance turns out to be approximately 65,000 kilometers, or ten Earth radii, although there is a probable range of from 50,000 kilometers to 80,000 kilometers, depending on the intensity of the solar wind at any given time.

If the tail of the magnetosphere were simply a sort of shadow cast in the solar wind by the forward part of the magnetosphere, then it would be a weak thing extending perhaps only as far as the Moon's orbit, which is at a distance of 400,000 kilometers, or 60 Earth radii. In fact the tail is much more substantial than this, apparently because the forward part of the magnetosphere is continually dragged along by the solar wind and the magnetic field is pulled out to great distances before being let go. This process results in the formation of a tail with a relatively high magnetic field strength and estimated to be perhaps 1 to 10 million kilometers in length. One theorist even claims that the tail extends as far as the solar wind blows, which is probably to the outskirts of the solar system several billion kilometers away. This, however, is to be regarded as a rather extreme point of view.

Near the Earth the shape of the geomagnetic field lines is similar to that of a dipole—that is, the field lines look as if they emanate from a small bar magnet placed near the center of the Earth. Field lines which reach the outer regions of the magnetosphere are quite different in shape, however, due to the distortion produced by the solar wind pressure and drag. One can understand how these shapes arise most easily by dividing the magnetosphere into two parts, the "doughnut" and the "tail." As its name implies, the former is a doughnut-shaped region which surrounds the Earth and extends to the magnetopause on the sunward side of the magnetosphere. The doughnut intersects the Earth (which occupies the hole) in two belts at about 20 degrees of latitude from the magnetic poles, corresponding roughly to the northern and southern auroral zones. Within the doughnut (i.e., at lower latitudes than the auroral zones), the magnetic field lines are dipole-like and link the two hemispheres. The magnetic field lines in the tail are in contrast quite undipole-like, and it is even questionable whether they run between the two hemispheres or instead are connected in some complicated way with the interplanetary magnetic field. The tail field lines meet the Earth in the polar regions at higher latitudes than the auroral zones, and extend out into the distant parts of the tail where they run approximately parallel to each other and to the solar wind.

The most interesting feature of the tail of the magnetosphere is that it is split along its length into upper and lower halves in which the directions of the magnetic field are opposite. If a compass needle were placed in the upper half of the tail, then it would point toward the Earth, that is, in the direction of a field line which leads to the geomagnetic north pole. If the same compass needle were carried into the lower half of the tail, it would swing round and point away from the Earth, that is, in the direction of a field line which leads away from the geomagnetic south pole. The two halves of the tail are separated by a relatively thin sheet in which the magnetic field direction reverses and the magnetic field strength is low; this is known as the "neutral sheet." The presence of the neutral sheet was first detected by a magnetometer flown on the satellite Explorer 18 in 1964, in an experiment conducted by Norman Ness of the Goddard Space Flight Center. The various other features of the magnetic structure of the magnetosphere which I have described here have been confirmed by numerous satellite observations, especially those from Explorers 10, 12, and 14.

I want to pass on now to discuss the properties of the plasma which is contained within the magnetosphere. But before doing so I must describe briefly how the motion of plasma is affected by the presence of a magnetic field. The most important point to remember is that the plasma and the magnetic field tend to be "frozen" together, in the sense that any two lumps of plasma which are at any time on a common magnetic field line always remain on a common field line. If, then, the magnetic field were very weak, it could be distorted and stretched into almost any configuration by movements of the plasma: this situation occurs in the interplanetary medium where the interplanetary magnetic field is combed out into a rather surprising spiral configuration by the joint effects of the solar wind and the rotation of the Sun.

On the other hand, if the field is relatively strong, it can exert a powerful controlling influence on the motion of the plasma: in fact, it can restrict the possible movements of the plasma such that the overall distortion of the magnetic field is minimized. This is the situation in the magnetosphere, for almost everywhere the geomagnetic field pressure greatly exceeds that of the plasma. Accordingly, the permissible motions of the magnetospheric plasma are such that the plasma on one field line interchanges with the plasma on adjacent field lines in such a manner that there is no net change in the configuration of the magnetic field as a whole. But despite the fact that the possible motions of the magnetospheric plasma are highly restrictive, it is significant that motion can take place. As I shall describe shortly, it is through such motions that the solar wind plasma can penetrate the magnetosphere, thus leading to the production of the aurora and the Van Allen radiation belts.

I have emphasized that the pressure of the plasma in the magnetosphere is almost everywhere small compared to that of the magnetic field. However, this is not the case in the neutral sheet in the tail. Here, in order to support the magnetic pressure associated with the higher magnetic field strength on either side, there must be plasma at an equal pressure to maintain equilibrium. This plasma presumably originates in the solar wind, which is able to enter the neutral sheet relatively freely from either side of the magnetosphere. According to current theories, the neutral sheet is the immediate source of the particles which produce the radiation belts. The injection takes place when the neutral sheet is for some reason thrown out of equilibrium, allowing the oppositely directed magnetic field lines on either side to link and contract and thus shooting plasma towards the Earth from the antisolar direction.

Within the doughnut-shaped part of the magnetosphere which surrounds the Earth, the plasma pressure is small compared with the magnetic pressure, but it is not entirely negligible. The plasma is in a sense "trapped" or contained by the magnetic field of the doughnut, which is in turn very slightly inflated by the small pressure exerted by the plasma. The inflation is observable as a very slight decrease of the geomagnetic field strength measured on the ground. Varying degrees of inflation of the doughnut, corresponding to changes in the pressure of the trapped plasma, can be monitored at magnetic observatories placed at various points around the world. The most pronounced fluctuations in the geomagnetic field strength observed on the ground have an amplitude of about 1 percent. These are called magnetic storms and are associated with violent auroral displays, ionospheric disturbances, and changes in the Van Allen radiation belts.

The plasma in the magnetosphere can be divided into two main componets. One of these is composed of low-energy particles, being essentially an extension of the ionosphere. The other consists of high-energy particles, most of which have presumably originated in the interplanetary medium beyond the magnetopause.

In the doughnut-shaped part of the magnetosphere where the geomagnetic field lines link the northern and southern hemispheres of the Earth, the low-energy component was detected first as a result of observations of a type of audiofrequency electromagnetic wave, usually called a "whistler." One can occasionally hear whistlers simply by connecting an antenna through an amplifier to a loudspeaker; the swishing tone gliding from high to low pitch gives rise to the name. These signals, which originate in lightning flashes, propagate from one hemisphere to another, approximately along the geomagnetic field lines, and sometimes have been observed to bounce back and forth many times before finally becoming too weak to be detected. This interpretation was given in the early 1950's by Owen Storey, then a graduate student at Cambridge University, who showed that it is possible to use the whistler observations to deduce the distribution of plasma density far out in the magnetosphere.

Perhaps the most interesting result which has been obtained in this way concerns the plasma density. Whistler studies have shown that the density of plasma within the doughnut does not decrease smoothly with distance from the Earth, but rather that there is a sudden drop which takes place on the field line which reaches out to about 20,000 kilometers. Little is known about the plasma in the tail of the magnetosphere, because the geomagnetic field lines do not connect the two hemispheres in this region and hence whistlers cannot bounce back and forth. It is expected, however, that the plasma density is low, for any plasma produced in the polar ionosphere by sunlight can drain off along the stretched-out geomagnetic field lines in the tail, and so be lost.

The high-energy component of the magnetospheric plasma completely fills the doughnut-shaped part of the magnetosphere and also the neutral sheet in the tail. Some of the particles constituting this component have energies sufficiently large to be detected with high efficiency by an ordinary Geiger counter, and it was these which were discovered in 1958 by James Van Allen and his colleagues at the University of Iowa. Prior to this it had been expected that only the background cosmic radiation would be observed by Geiger counters carried into space; it was a great surprise to find particle fluxes of such intensities that the first instruments were saturated.

In the early experiments a belt of energetic protons was found to encircle the Earth's equator out to a distance of several thousand kilometers. This is now known as the inner Van Allen radiation belt; it has the property of being remarkably stable, and in deed only recently has it been shown by careful observation that slight variations do occur. Subsequent experiments at greater altitudes showed the existence of a second belt of energetic particles which extends out to about 20,000 kilometers above the Earth and which became known as the outer Van Allen belt. The particles in this case are electrons and, in contrast to the inner belt, the fluxes observed are quite variable with changes related in some complicated way to magnetic storm effects.

Surprisingly enough, although the Russians were first to put scientific satellites into orbit, they did not observe the radiation belts until later. The reason for this was that the first Sputniks orbited at relatively low altitudes, where the residual atmosphere is sufficient to remove any of the radiation-belt particles that might penetrate that far. The first scientific satellite launched by the United States, Explorer 1, went into a much higher orbit and was therefore favorably placed for observing the radiation belts. One can detect from hindsight some trace of the radiation belts in the Russian data, but this would have caused no interest if it had not been for Van Allen's observations, and it is to him therefore that the credit for discovering the radiation belts properly belongs.

Since 1958 a great number of observations have been made from Earth satellites and space probes, and it is now known that the whole of the doughnut-shaped part of the magnetosphere is filled with energetic particles. Van Allen's relatively insensitive instruments found at first only the more easily detected particles, which are as deceptive a representation of the whole population as is the visible portion of an iceberg. The various components of the radiation belts, as observed by different types of detectors, have widely different characteristics; hence it is rather difficult to make general statements about the energetic particles as a whole. However, one thing is certain: since these particles contribute the bulk of the pressure exerted by the magnetospheric plasma, any inflation of the magnetosphere that takes place must be associated with an enhancement of the total energy in the radiation belts. The magnetosphere is observed to become inflated during magnetic storms, and at such times we can be sure that the radiation belts have been enhanced, even though some particular groups of particles might seem to indicate the opposite.

Although they were a little unlucky in failing to make the first observation of the radiation belts, the Russians did discover in 1959 what at the time they called the third radiation belt. The discovery was made by means of a charged particle trap carried on the Moon probe Lunik II, and was due to K. I. Gringauz, who also first observed the solar wind directly in a similar experiment carried on Lunik III. We now know that the radiation belt observed by Gringauz is in fact associated with the neutral sheet in the tail of the magnetosphere rather than with the doughnut. These particles, which are probably relatively new to the magnetosphere, seem to originate in the solar wind, and are believed to be on their way into the doughnut where they feed the Van Allen belts and also give rise to the aurora as they precipitate into the atmosphere.

To conclude this discussion of the Earth's magnetosphere, I want to describe very briefly the magnetic-storm phenomenon. The cause of magnetic storms is a long-standing problem of much interest, but prior to the space age little progress had been made towards its solution, despite the fact that many aspects of the phenomenon can be easily observed from the ground. I am referring, of course, to the geomagnetic fluctuations, which were first observed 160 years ago by von Humboldt, to auroral displays which have probably been known throughout recorded history, and to the ionospheric disturbances which interfere with long distance radio transmissions, especially in the polar regions. Our knowledge and understanding of magnetic storms has grown enormously since 1958 as new discoveries have been made in space, and at this stage the solution to the problem seems to be almost at hand.

The sequence of events constituting a magnetic storm typically begins with the occurrence of an explosion (or flare) on the visible hemisphere of the Sun. The material ejected by the explosion blasts its way through the interplanetary medium at a speed of 1,000 to 2,000 kilometers per second. Nothing happens at the Earth until a day or so after the flare, when the front of the blast envelopes the magnetosphere. The sudden increase of the external pressure at this point crushes the magnetosphere and causes the geomagnetic field strength observed on the ground to increase. This increase is maintained for several hours and is known as the "initial phase" of the storm. In addition to squeezing the magnetosphere during this period, the enhanced solar wind accompanying the blast drags the outer parts of the magnetosphere along with it, thus causing the tail to grow at the expense of the doughnut. Eventually, however, the growth of the tail is arrested by its lack of stability. This apparently comes about because the plasma supporting the neutral sheet is unable to prevent the magnetic field above and below the sheet from linking to form closed but extended loops.

The newly linked field lines in the interior of the tail contract violently, carrying plasma toward the Earth and into the doughnut-shaped part of the magnetosphere on the night side. The contraction of the field lines is observed on the Earth as a severe magnetic disturbance in the polar regions lasting about an hour, and called a "polar substorm." This is accompanied by an intense auroral display produced by the precipitation into the atmosphere of part of the plasma that has been carried in from the neutral sheet. The remainder of the plasma is injected into the doughnut, which becomes partially inflated. The entire doughnut becomes inflated after about an hour or so, causing the geomagnetic field strength on the ground to decrease to less than its prestorm level; this is called the "main phase" of the storm. Several polar substorms might occur if the enhanced solar wind continues to enlarge the tail, and each one leads to further inflation of the doughnut. Eventually, however, things settle down and we are left with the doughnut inflated, and correspondingly an enhancement of the total energy of the radiation belts. The inflation, together with the associated depression of the geomagnetic field observed on the ground, subsides slowly as the newly injected plasma leaks out of the magnetosphere—either into the interplanetary medium or into the atmosphere. This recovery phase lasts several days.

This description of a typical magnetic storm contains a judicious blend of fact and theory and is all rather qualitative. However, there is fairly wide support for the views I have advanced and I have little doubt that when the full story is known, they will prove to be not too wide of the mark. Nevertheless, these are exciting times for space physicists, with major new discoveries being made almost monthly, and one can be sure that there are many surprises still in store for us.

• 13

The Earth from Space

W. M. KAULA

The subject I am concerned with is geodesy: the application of the most familiar mathematics and the most familiar physics to a very familiar object, the Earth. The mathematics is essentially the geometry of three-dimensional space associated with the name of Euclid. The physics is that of gravitation, which was largely founded by Newton: most important is Newton's law stating that two masses will attract each other proportionately as the product of their masses and inversely proportionately as the square of the distance between them. We are interested in applying this Euclidian mathematics and Newtonian physics to the Earth: to determine what its shape is and how big it is: its size in terms of linear dimension, as well as its mass in grams or some other such unit. We are also interested, of course, in determining variations in the distribution of its mass and variations in its shape.

It took mankind many centuries of historical time to agree that the Earth is rather round. Some of the ancient Greeks, such as Eratosthenes, believed the Earth was spherical. Eratosthenes made a pretty good measurement of the Earth's size by using the difference between the lengths of shadows cast by the Sun at Alexandria and at a point some hundreds of kilometers south. We also all know that the Earth pulls on any object with an acceleration that is the same for all objects, as was first demonstrated by Galileo, allegedly using the leaning tower of Pisa as his platform. This acceleration is what we call "gravity."

Besides the acceleration of a dropping object, there are other things which are affected by gravity, such as the pendulum which, at the upper limit of one swing, has an acceleration downward which causes it to acquire a velocity which reaches its maximum at the midpoint. This velocity carries the pendulum up on the other side until the counteracting acceleration of gravity cancels it out. The cycle is then repeated with a period which is dependent upon gravity. We can also measure gravity by a spring balance: if we attach a mass to the end of a spring, the pull it will exert upon the spring depends upon the gravitational acceleration.

If the Earth were fluid and only the law of gravitational attraction applied, the Earth would be a perfect sphere. However, since Newton, more detail has been deduced about the shape of the Earth. Newton added the element of centrifugal force: as a body rotates, it will tend to throw things out, in opposition to the centripetal attraction of gravitation. Because this outward force is greater at the Earth's equator than at the poles, the Earth is pulled outward at the equator and has a flattened shape; consequently, at the pole we are about 21 kilometers closer to the center than we are at the equator.

In addition to the various scientific investigations applying the mathematics of Euclid or the physics of Newton to the Earth, there has, of course, been much more in the way of practical applications. A familiar technique is surveying, which has been developed over several centuries for the purposes of engineering construction and of making maps for navigation, delimiting property boundaries, fighting wars, and so on. The efforts of many surveyors over many years have resulted in a system of measurements of the Earth which is a concatenation of lengths measured by tape and of angles measured by a combination of a telescope and a graduated circle (the combination is called a theodolite). These networks of overlapping triangles have now been extended so that, for example, in the western hemisphere there are connections all the way from Alaska down to the southern part of

Chile, and in the eastern hemisphere from Lapland down the Cape of Good Hope in a north-south direction, and from Great Britain to Japan in an east-west direction. These survey efforts measure the distances and directions between points on the Earth's surface; used in conjunction with observations of the directions of a plumb line with respect to the stars, they determine variations in the direction of the pull of gravity.

There has also been considerable effort, entailing a similar concatenation between reference points, at measurements of the intensity of the acceleration of gravity by pendulums and by spring balances very highly refined compared to those with which we are familiar at the grocery store. From these systems of measurements we have deduced a lot more about the shape of the Earth than that it is simply the flattened sphere or, more strictly speaking, the oblate ellipsoid of revolution which Newton deduced would be the shape of a fluid body under the combined influences of gravitational attraction and centrifugal acceleration due to rotation. Therefore it is appropriate to define much more precisely what we mean by the shape.

The Earth has a rather fuzzy outer limit, and if we are to have a reasonably precisely defined shape, we will ignore that one part in a million of the Earth that is constituted by its atmosphere. This leaves us with two obvious alternatives as the Earth's outer limit: the surface of the rocks and the surface of the sea, sometimes called the lithosphere and the hydrosphere. (The lithosphere is the surface whose variations are most important to us because if it did not exist we would all still be fish.) We commonly regard the heights of the mountains as an excess, a surplus of matter, and the depths of the oceans as a deficiency of matter. But during the nineteenth century, as more and more measurements were made of the direction of gravity by the combination of triangulation and astronomical observations and of the intensity of gravity by pendulum measurements, it was found that the effect of mountains, such as the Himalayas, on the direction and intensity of gravity was not as much as was calculated from the size and shape of the mountains and the density of the rocks.

This phenomenon was first noticed by the French geodesist Pierre Bouguer in the Andes in the eighteenth century. It has since been found to be true over most of the world that, where there is a large excess of matter at the Earth's surface, such as a mountain range, it appears to be compensated by a deficiency of mass somewhere down deeper in the Earth; and that, conversely, where there is a deficiency of mass, such as an ocean basis, it is compensated by an excess within the Earth. This "compensation" appears to occur at depths of the order of a few tens of kilometers. The arrangement of the crust suggests roughly the situation of a floating iceberg: the iceberg has a mass deficiency of the portion below the water surface which exactly compensates for the mass of the small part which extends above it. This balancing of excesses and deficiencies is known as *isostasy,* and is a fundamental characteristic of the Earth's crust.

Because of the existence of isostasy, it is therefore more meaningful to select or define, as the external shape of the Earth, not the surface of the rocks but something which is more expressive of the distribution of matter or mass: in other words, a surface defined by a gravitational pull. The simplest surface defined by a gravitational pull is called an "equipotential." An equipotential is a surface which is everywhere normal, or at right angles, to the acceleration due to gravity. Examples of equipotentials are the surface of the water in a pond, in a bathtub, or in an ocean. Because the ocean falls into this category, the most obvious choice of an equipotential would be the sea surface. The sea surface goes up and down, owing to tides from the Sun and the Moon; so we select the average, the mean sea level. Manifestly, gravity does not stop at the border of the ocean but continues on land, and hence we continue inland this same mean-sea-level surface at right angles to the gravity acceleration. When using the mean sea level in this manner, it is normally referred to as the "geoid."

As we extend the geoid through the continents, there is the problem of some mass outside the geoid: the land above sea level. This mass would affect the gravity acceleration, in turn affecting the geoid. Because we do not know the density of the rocks exactly, we cannot determine the geoid exactly. The refinement in definition of the geoid entailed by this difficulty is a subject of considerable debate amongst the more mathematically inclined geodesists; however, the differences which could reasonably exist are still a good deal smaller than the accuracy with which we can deduce the geoid from existing measurements. The important fact is that if we could extrapolate the relatively short distances from the geoid to an equipotential which completely evelopes the significant mass of the Earth, and then completely define the shape of this enveloping equipotential, we would thereby know completely and exactly the gravitational acceleration of the Earth at every point outside the Earth throughout space.

Having refined the definition of shape, let us return to the concept of isostasy. Isostasy, the balancing of excesses by deficiencies and vice versa, is far from perfect; as a working tool, it is significantly imperfect. It is obvious that over short distances isostasy does not apply: a small hill a few kilometers in extent is an excess which is small enough to be sustained by the crust. But despite these localized variations isostasy generally prevails on a regional scale, say, of several hundreds of kilometers. However, how much it prevails on a long-range scale of thousands of kilometers—the scale of ocean basins or major geological provinces of the continents, etc.—has long been a matter of debate. This debate is caused mainly by the insufficiency of the data, which gives rise to a variety of plausible interpretations and hence to a very wide range of opinion as to the size of the variations of the geoid on a global scale. What was needed was some way to stand back from the Earth and to take an overall look, some better way to deduce these variations than from analysis of gravimetry measurements, which included a very large clutter due to local variations, such as the ups and downs due to hills and other small features.

The obvious device came along a few years ago in the form of artificial satellites which travel around the Earth in orbits determined by the Earth's gravity field. The orbits of the planets and satellites were, of course, the principal data which stimulated Newton to hypothesize his law of gravitational attraction. For geodesy, satellites can be regarded as falling objects whose paths of fall are used to measure the Earth's gravity field. If there is an object which is given a certain velocity in a horizontal direction, we know that it takes time for the acceleration of gravity to give the object a specified downward velocity. In that time it will have gone forward to some extent. Now if the object has enough forward velocity, then by the time it has, so to speak, "dropped," it will have gone "around the corner" of the Earth, and this velocity of dropping will be a horizontal one that will cause it to keep on going around. This "going around" can keep on going forever, provided the object was given enough initial velocity.

If the Earth were a point mass or a perfect sphere, the path the object would follow would be an exact ellipse, with the Earth at one focus. At the lowest point, called the perigee, we would have a maximum velocity, which would be in excess of that required to keep it at the same altitude, so it would rise. In rising, the satellite would lose some of its velocity against the Earth's pull until it eventually reached apogee, the point of maximum altitude but minimum velocity. The satellite would not have enough velocity to stay at apogee altitude, so it would fall, and in falling again increase its velocity so that it would return to the same point of maximum velocity, the perigee. The process is a continual interchange between stored potential energy and active kinetic energy, somewhat similar to that of the pendulum.

The thing which makes satellites interesting, however, is that the Earth is not a point mass. There is firstly the oblateness, the flattening due to the Earth's rotation, whose effect is to exert a sideward pull on the satellite. A sideward pull on an object in rotational motion, such as a top, produces a precession, a gradual motion of the axis about which that rotation is taking place. Consequently, the dominant difference of satellite orbits from the central field ellipse is a precession of that ellipse on the order of a few degrees per day due to the flattening of the Earth. However, there is much more involved in the motion of a satellite close to Earth.

Satellites are observed to waver gradually back and forth around this precessing ellipse. These wavers or wobbles amount to as much as 5 kilometers. Observations of satellites by photographing them against the stars with large cameras or by listening to the Doppler signals from a radio beacon on the satellite are quite accurate: they are sensitive to variations in the satellite positions of the order of a few meters. So we should expect that, if a satellite is close enough to the Earth, we could determine from it variations in the Earth's field which cause wobbles of a few meters or more. Because we know from terrestrial gravimetry that the variations are of the order of a few parts in a million, we should expect in the course of a day that the satellite's path would vary by a few parts in a million from the distance which it travels in a day (about 500,000 kilometers). Hence, we should expect it to vary by several hundred meters a day.

Let us now take this satellite orbit and regard it as the sum of two parts: first, a perfectly precessing ellipse plus, second, the variations of the actual orbit around that ellipse—the wavers and wobbles I mentioned. Next, if we take the perfectly precessing ellipse and hypothetically unwrap it into a single line, each point on this line will correspond to a certain point in time, for the satellite is always on a certain point on the precessing ellipse in time. We can then superimpose on this time line the second part of the orbit—the wobbles and wavers. Looking at the orbit in this way, the wobbles and wavers constitute what is called a "time series," similar to those used to characterize radio wave propagation, the tides, or the ups and downs of the stock market.

Like any time series, this series can be broken down into what is called the spectrum: the irregular, or the more-or-less irregular, time series can be considered as the sum of several simpler and more regular curves which are sinusoidal curves, each with its own fixed wavelength in time. More commonly used than wavelength to define one of these curves is frequency, the inverse of wavelength. In the case of a satellite orbit, the spectrum comprises frequencies of the order of some cycles per day (depending on the rotation of the Earth with respect to fixed space), or even of the order of a few cycles per year (depending on the rate of precession of the orbit due to the flattening of the Earth). Corresponding to certain frequencies in the orbital variation in time, there are certain variations of the gravitational field in space. If we use several different satellite orbits and lots of observations of them, we can distinguish these variations in the Earth's gravitational field as different wave components. These waves are similar mathematically to the waves of the time series in that the irregular total can be regarded as the sum of several different regular components of differing amplitude and wavelength.

However, the Earth's gravity field is different in that it is spatial rather than temporal: it is fixed in time but varies in space over the Earth's surface; moreover, instead of the field being along one axis as is the case with time, it is a two-dimensional continuum, the Earth's surface.

Now, from satellites and the observations of them over the last few years, we have been able to establish, in effect, regular oscillations of the orbits with an amplitude of the order of about 5 meters and from these, in turn, to deduce several dozen of the different terms in the two-dimensional spectrum of the Earth's gravity field. From such analyses, the picture we get is the shape of the Earth which we previously defined as the geoid: the mean sea level as continued through the continents. We can draw a map of this geoid, much like the familiar map of the continents.

However, this map looks somewhat different from the continents, the rocky surface of the Earth. It has its ups and downs in quite different places. The most marked ups and downs happen to be in the eastern hemisphere. The dominant feature of the eastern hemisphere is a hollow which is centered slightly southwest of the southern tip of India. Referring the equipotential of the Earth to the best-fitting ellipsoid of revolution, this Indian Ocean minimum appears as a hollow about 90 meters deep. This hollow extends up northward across the Himalayas, resulting in the paradox that the maximum excess in the topography just happens to be part of a deficiency in the gravity field. Around this great Indian hollow in the geoid, there are three maxima: one to the east, centered over New Guinea, about 75 meters high; one to the southwest, below the Cape of Good Hope, about 50 meters; and one to the northwest in the North Atlantic just off Great Britain, about 60 meters. There are smaller variations, of course, which cause oscillations and slight saddles in the shape of the geoid around these four dominant features of the eastern hemisphere.

In the western hemisphere the geoid variations are not so pronounced. The dominant features are two hollows of about minus 50 meters: one in the Atlantic off Florida and one in the Pacific off Lower California. The saddle in between these hollows is about minus 20 meters. Further south, there is a peak in the Andes around Peru of about plus 40 meters. Finally, there is a minimum down in Antarctica on the Pacific side of about minus 50 meters. The saddle between this Antarctic minimum and the minimum is the Pacific off North America is about minus 10 meters. We thus have a fairly interesting-looking picture in which the dominant features, spaced at distances of about 60 degrees of arc, have no apparent relationship to the continents and oceans.

The scientific interest of this geoid picture is that the corresponding variations in the gravitational attraction imply that there must somewhere be variations in density within the Earth, which in turn entail some type of stress difference. Density irregularities cause variations in the attraction of the mass of the Earth for the irregularities, in turn causing variations in pull on the rocks which tend to break the rocks in shear. A part of this shearing stress could be relieved by a much deeper isostasy than that associated with the regional variations of the gravity field which I previously mentioned. There could exist a balancing of positive and negative anomalies which would result in no stress below a depth of, say, a few hundred kilometers, which would still account for the variations shown at the surface. However, such an elaborate, deep-seated isostasy would imply some rather complicated scheme of fractionation of materials in the Earth and convection, or movement of matter with heat. On the other hand, if the irregularities are supported by shearing-stress differences, the strength of the materials within the Earth require to support these stresses statically is difficult to reconcile with the strength of rocks under high pressures and temperatures as observed in the laboratory.

The figure to which we referred our geoid was that of a best-fitting ellipsoid of revolution—a mathematical fiction which is convenient for calculation. If we were to select a geophysically more meaningful reference figure, it would be the shape of a rotating fluid, with the same mass, radius, moment of inertia, and rate of rotation as that of the Earth. This shape differs from the one that best fits the Earth in the mathematical sense by something of the order of twice any other variation in the gravitational field. In other words, if you use such a reference figure, our geoid map would be mainly a north-south variation, and the dominant feature would become the maximum over New Guinea rather than the hollow in the Indian Ocean.

But the fact that the flattening of the Earth is associated with the rotation of the earth suggests a special explanation. If we were to turn the calculation around and take as fixed the observed flattening, and leave as an unknown the rate of rotation, we would get a rate of rotation which was about three parts in a thousand higher than the actual rotation. If we combine that with the rate of decrease of the Earth's rotation observed from the disparity from the Newtonian theory of the motion of the Moon due to tidal friction, we would conclude that the present shape of the Earth is what it should have had about 15 million years ago. And so, because this term is extra large, we perhaps are right in giving it a special explanation: namely, that it is a measure of the delay on the part of the Earth in compensating its shape for the slowing down due to tidal friction with the Moon.

To explain the other variations we must look elsewhere, to other geophysical information, and determine which of these appear to have some relationship to gravity. Of the various other things which are measured with regard to the Earth, the one which seems to be most readily comparable is the measurement of the flow of heat out of the Earth. This is a very small amount, about 1/25,000 of the heat we receive from the Sun, but is still quite perceptible. Its source is most likely to be the radioactive decay of uranium, thorium, and potassium within the Earth.

Heat flow shows quite a definite negative correlation with the geoid: there are maxima in the heat flow where there are minima in the geoid, and vice versa. The simplest explanation is that hotter rocks are of a lower density due to thermal expansion and, conversely, colder rocks are contracted. Whether there is a further correlation due to the conveying of heat to the surface by convection currents is a question which is difficult to answer because of the present inadequacy of our theory to cope with such complicated systems of motion.

Another clue which bears on the possibility of convection currents is—on a time scale of the order of a few thousand years—the observed rate of uplift of areas in northern Europe and North America, from which a load has been removed in recent geological time by the melting of the icecaps. If we use these rates of uplift to deduce the viscosity of the Earth (the rate at which it moves in response to stress), we find that the material in the Earth would be moving at the rate of a few centimeters per year in response to the irregularities in the density which correspond to the gravity field. This rate is also that which is deduced for a much longer time scale—of the order of many millions of years—from the apparent variation in the direction of the north pole for different continents as deduced from residual magnetism: the magnetic orientation of iron-containing minerals in lava which poured out onto the Earth's surface and then cooled below the maximum temperature at which iron can be magnetized. There are various other indicators of activity in the Earth, such as the properties deduced from seismic waves from earthquakes. Most of these indicators suggests that the upper mantle, the section of the Earth which is 50 to 400 kilometers deep, is weaker and more active than the lower part of the mantle. However, we still do not know the sources of the broad-scale variations of the gravity field: whether they are the result of slow convection in a weak upper mantle or whether they are broad-scale variations which have been frozen into the lower mantle since its creation early in the Earth's history some billions of years ago. But the current rate of improvement in our knowledge of the gravity field and other geophysical quantities is such that we can reasonably hope to deduce a much more accurate model of the behavior of the Earth's interior, and hence its past history.

14

The Moon

HAROLD C. UREY

Although the Moon has been an object of interest to men from the earliest historical times, it is nonetheless surprising to realize how much the ancient Greeks knew about it. Anaxagoras, who lived from 500 to 428 B.C., understood what caused the eclipses of the Sun and Moon and Aristotle, who lived from 384 to 322 B.C., recognized that the Earth must be round because of the circular character of the eclipses of the Moon. He also realized that the same face of the Moon was always turned toward the Earth. These ancient observations were lost to men for approximately two thousand years until Galileo in 1610 invented his small telescope and recognized that the Moon had craters and mountains upon it. Studies since that time have elucidated the structure of the Moon as it can be seen in telescopes from the Earth.

At first it was thought that the great dark areas of the Moon were seas and the bright areas were continents. Hence, the dark areas were named "maria" (Latin for "seas"). It was also thought that the mountainous areas were mostly of a volcanic origin, a view held until the latter part of the nineteenth century. Since then we have realized that most of the features of the Moon have been produced by collisions of objects with its surface, though there are some features that must be regarded as volcanic in origin. This perhaps was most clearly stated by G. K. Gilbert, an outstanding American geologist, who studied the Moon during the

1890's. That collisions must occur with the surface of the Moon is now well established. We know that small objects such as meteorites fall on the Earth and we also know that considerably larger objects capable of producing large collisional craters have fallen on the Earth: the record can be found and can be studied. Also, we know that comets should collide occasionally with the Earth. Of course, if these objects collide with the Earth, then they must also collide with the Moon. It is very likely that all of the larger craters of the Moon as well as most of the smaller ones are indeed due to collisional effects of this kind either during the long time—approximately 4.5 billion years—that the Earth and solar system have stood, or perhaps due to collisions during the early history of the solar system, that is, during the time that the Earth was accumulating out of objects of this kind and growing to its present size.

The origin of the maria is not quite so certain. Some of these great smooth planes are nearly circular in outline, and in these cases we strongly suspect that collisions of larger objects with the Moon produced these supercraters which then became filled with smooth grey material of some kind. Whether the craters and the maria are filled with lava, as many people think, is not so certain as one might gather from popular discussions. It is surprising that lava flows should have occurred so generally over the surface of the Moon as to fill up all the regions within craters, between craters, and so forth, with lava. Some of us feel that these smooth areas are filled partly at least with fragmented material of some kind or other. On the other hand, these smooth areas of the maria may be filled with lava, that is, rocky material that has been melted. Recently it has been pointed out that there are certain smooth areas that are comparatively free of the many small craters that are so characteristic of other areas, and this indicates that some recent event has produced these smooth areas in limited regions of the Moon, which argues for some sort of lava flow. There are also marked differences in the color of the smooth areas of the Moon which can be quite easily interpreted as due to the flow of lava over some areas and not over others, or a succession of lava flows with flows that have occurred at different times having somewhat different colors, that is, some more gray than others.

One should always keep in mind that small planetary objects

such as the Moon should, in general, show less volcanic activity than large objects such as the Earth, for we strongly suspect that the origin of the melting is due to radioactive heating, and the amount of radioactive heat produced in an object is proportional to its volume, which is in turn proportional to the cube of its radius. But the rate of loss of heat from the object will be proportional to its surface, that is, to the square of its radius, and hence one expects the larger object to be hotter and to produce more extensive volcanic activity and lava flows. Of course, it is not necessarily true that the history of the Moon is similar to that of the Earth. Some other sources of energy (which the proponents of extensive lava flows do not generally trouble to explain) may have produced volcanic activity in excess of that expected for a small object such as the Moon.

Then again the surface of the Moon may be covered with dust, finely fragmented material, and things of this sort. Several sources of material of this kind have been suggested. In the first place, one must expect that the great collisions that have quite certainly been part of lunar history should have produced great clouds of finely fragmented material. Possibly these collisions would also result in the presence of a temporary atmosphere which would lead to a dispersal of dust over great areas of the Moon, perhaps in a rather smooth blanket such as we see for great areas of the Moon. But on the Earth too we have enormous ash flows which are a part of the volcanic activity of the Earth, and it has been suggested that some of the great smooth areas are, indeed, the analogue of these dust flows on the Earth. Finally, it has further been suggested that the great collisions may have released large amounts of finely divided material which flowed out of the crevasses produced by the collisions, that is, again a dust flow but one due to quite a different physical source.

Quite regardless of what the origin of the materials of the surface of the Moon may be—fragmented material, lava flows, and so forth—it seems fairly certain that there is a layer on the surface of the Moon consisting of dust due to the effects of collisions with the surface: namely, meteorites large and small, microscopic meteorites, the effect of high-energy particles from the Sun, and cosmic rays—and aided by heating and cooling of the surface by the Sun. All of these processes should have produced some thickness of fragmented material regardless of what the original material of the lunar surface was.

A great question arises as to how thick the layers of dust of this kind might be. The optical properties of the lunar surface are those of what are called "fairy castle" structures—many finely divided little particles forming a loose structure so that the light entering the openings will be reflected several times before it emerges again. But it is difficult to say how thick this layer is. So far as the light observation is concerned, it could be very thin, say, a fraction of a centimeter in thickness. Some people have argued that it is a completely bare surface. This I think cannot be true. There must be at least a sufficient thickness of dust to account for these optical properties. In discussing the Ranger and Surveyor pictures below we will return to this point, but I would anticipate that discussion by simply saying that it is my conclusion that there is nothing in these pictures which definitely decides this question of how thick the layer of fragmented material may be. Some laboratory experiments certainly indicate that there is a considerable layer of fragmented material on the surface of the Moon.

The question of the history of the Moon is a most intriguing one which, of course, we cannot follow except in a deductive way and oftentimes in a highly mathematical way in accordance with physical laws. At present the Moon is receding from the Earth. This has been definitely established by the dates and places where eclipses of the Moon were recorded in ancient times. Because of the high precision of astronomical observations and the exactness of the laws of mechanics, it is possible to predict exactly when and where on the surface of the Earth eclipses should have been observed. We find that there are discrepancies in the historically recorded eclipses, particularly with respect to their positions on the Earth and, because the Earth rotates, the discrepancies mean that the times of the eclipses were different from those calculated.

As a result of this we conclude that the Moon is moving away from the Earth, and if we set up our mathematical formulas for this, as has been done, we find that the Moon should have been near the Earth some 1,500 or 2,000 million years ago, whereas our dating of meteorites indicates that the origin of the solar system occurred approximately 4,500 million years ago. One can only wonder where the Moon was for this roughly 3,000 million years. There appears to be no satisfactory place to have stored it. It seems likely that the tidal effects have been different in the past from what they are now and that the Moon did not recede from the Earth as fast as it is receding now. Hence, probably, the Moon originated at about the same time that the Earth did. Probably the Sun, the Moon, the Earth, and all the planets originated 4,500 million years ago. This date has been definitely determined for the meteorites, and it is very probable that they acquired their mineral structures during the time that the Earth and planets were being formed.

Several ideas have been presented as to the origin of the Moon. During the last century, Sir George Darwin studied the Earth-Moon system with great care, and he proposed that the Moon separated from the Earth due to tidal action. This idea of the separation of the Earth and Moon is not generally accepted at the present time but other suggestions have been made. It is supposed that the high-density core of the Earth was distributed throughout the Earth. Differences in the distribution of the iron of the core result in a change in what is called the moment of inertia of the Earth. The moment of inertia is secured by multiplying the mass at each point in the Earth by the square of its distance from the center and adding these quantities all together. If the Earth has uniform density throughout, this value should be 0.4 times the mass multiplied by the square of the radius of the Earth, whereas this quantity for the Earth at the present time is only 0.334 times the mass times the radius squared. Now, if this moment of inertia is multiplied by the angular velocity, it should give us angular momentum, which is a constant according to the mechanical laws of Newton. Hence, if the moment of inertia should become smaller, the velocity of rotation must increase, and it has been suggested that the formation of the Earth's core increased the velocity of rotation and some material was lost from the Earth which formed the Moon. Most competent students of this subject believe this process is not possible and that the Moon could not have originated in this way, but it is very difficult to get general agreement on this subject. Perhaps the Moon separated from the Earth. I personally doubt that this conclusion is correct.

The second suggestion is that the Moon accumulated out of solid objects in the neighborhood of the Earth. If this is the case, it is necessary to account for the difference in composition of the Earth and Moon. The Earth must contain something like 30 percent by weight of the element iron in order to account for its density, whereas the Moon contains 10 to 15 percent of iron on the basis of its density. It is necessary in this case to account for the difference in density of the two objects. Why did iron accumulate in the Earth to a larger extent than it did in the Moon? No very satisfactory explanation has been given, and this does not seem to be a very likely method for the origin of the Moon.

The third suggestion holds that the Moon was captured by the Earth. It is very difficult to understand how the Moon could have been captured by the Earth. The Moon would approach the Earth in what we call a hyperbolic orbit, pass somewhere near the Earth, lose a little energy due to tidal effects, and just barely not escape, and it would then be moving in a very long elliptical orbit. The question is: How did the orbit get rounded up into a nearly circular one? One might speculate that the Moon collided with objects in the neighborhood of the Earth in order to accomplish this, and possibly something was added to the outer part of the Moon in this process. But of course in this case we must expect that the outer part of the Moon would have a composition similar to that of the Earth. We must account for a Moon being formed out of comparatively low-density material and then acquiring a layer of some thickness on the outside of high density material like that of the Earth. This leads to a complicated model for the origin of the Moon, and we might say that no method for the origin of the Moon is possible and the Moon simply cannot exist—but there it is, just the same.

Possibly we will find out more about this when we get samples back from the Moon. Possibly the material accumulated on the surface and then sank to the interior, having a higher density. But we must then also account for a Moon of low density captured by an Earth of high density. It is very interesting that the composition calculated for the Moon with respect to the rocky materials and iron is more nearly like that reported for the Sun with respect to the same elements rather than that for the Earth or other terrestrial planets. It looks as though objects having nearly the composition of the Sun may have been important in the early history of the solar system. But it should be emphasized that none of these conclusions is certain. Of course, if we were certain about all of these things, there would be no particularly good reason for investigating the Moon. I have a prejudice in regard to this subject. I should like to have the Moon be interesting. I should like to have it tell us something about the early history of the solar system and I rather think that it will.

The National Aeronautics and Space Administration has succeeded in sending several Ranger and Surveyor missions to the Moon. These took pictures which were televised to the Earth. The pictures give us much greater detail in regard to the surface than any terrestrially based photographs. The three successful Ranger missions landed in the Oceanus Procellarum, the Mare Serenitatis, and the Alphonsus crater. All of these are in smooth areas of the Moon rather than mountainous ones. One reason for selecting these sites was that we would like to know where the Apollo mission, which will carry men to the Moon, might land and what the conditions would be. Also, the Alphonsus crater shows halo craters in terrestrial photographs. (A halo crater is a small crater surrounded with a small black patch. It is quite certainly of plutonic origin and looks as though gaseous explosions from the crater had thrown material above the crater and dropped it in a little patch in the neighborhood.)

The facts learned on these missions are not markedly different from those deduced from the study of photographs of the Moon taken with large telescopes. We still think that the big craters on the Moon are due to primary collisions of objects with the surface of the Moon. It seems doubtful that they are due to sinking of great areas of the Moon into cavities below the surface. Also, there are secondary craters produced by the fall of objects resulting from the production of big craters. Thus we find secondary craters that were produced by objects thrown from the crater Tycho or the crater Copernicus in the Ranger 7 pictures. We find craters of this kind in the Ranger 8 pictures as well. Possibly the new feature shown by these pictures is that certain craters look very much as though they are collapse features on the moon where areas have sunk below the surface. These are not dissimilar to collapse features seen in volcanic areas of the Earth, but mostly those on the Moon are very considerably larger than those on the Earth and hence probably have a different origin. Sometimes the collapse features take the shape of small dimples of a rather symmetrical kind, as though finely grained material sank through little holes at the bottom somewhere and left a dimple on the surface. Sometimes they look almost like a funnel with rather straight funnel-shaped walls, and sometimes they are much more indefinite than this, as though there was some irregular collapse below the surface. Certain of the larger craters as seen from land-based photographs have rugged walls with collapse features on the interior walls, having rather rugged irregular shapes oftentimes with a central peak. But, also, certain of these craters are very smooth with smooth walls. This indicates a different origin for the two, and the smaller craters observed in the Ranger pictures show both types of craters.

On the Earth we have been able to make experiments with high-velocity objects falling into material of different kinds. It is found that in order to produce the smooth well-shaped craters our projectiles should land in finely divided unconsolidated material like sand, whereas missiles landing in solid material having considerable physical strength make irregular-shaped craters and always throw out some objects of considerable size. It is also found that ordinary explosives or atomic bomb explosives detonated in material of physical strength throw out secondary objects of sizes that are rather proportional to the size of the craters. These experiments would seem to indicate that there is a layer of material on the surface of the Moon that is probably some tens of meters of thickness of rather poorly consolidated material underlain with material of very considerable strength. It would seem that this would account for what is observed, though it is difficult to scale up small experiments and draw valid conclusions in regard to much larger events such as are seen on the Moon. Collapse features are observed in volcanic lava flows on Earth, but those on the Moon are very much larger and are probably of different origin.

The Ranger pictures have not really resolved the problem of the fundamental materials of the lunar surface. They did not decide whether it is highly fragmented material, dust, dust flows, or lava. They did not show what the strength of the material of the lunar surface is, and they did not decide whether it is sufficiently strong to support the weight of the Apollo vehicles. The Ranger pictures have been very fascinating but they did not tell us much beyond what we had deduced from terrestrially based pictures.

The Russians have recently sent two interesting vehicles to the Moon: Luna IX and Luna X. Luna IX sent a capsule which soft-landed on the Moon and took a limited number of pictures. On its soft landing it did not sink very deeply. It scanned pictures by a rotating device that attempted to take in the whole horizon. It was tilted somewhat, so part of the horizon was missing; but after it scanned around and came back to its previous position, it had shifted and tilted to a different angle. This would seem to indicate an unstable position on the Moon, as though it had landed on rather soft, unsubstantial material. It took pictures down to small sizes, even millimeter-size objects, and showed rocks and pebbles and craters on the lunar surface. In some cases it looked as though material had been eroded away and left little pedestals sticking up above the surface capped with some resistant material. One wonders where the eroded material went to. Was it thrown off into space, or did the finely divided material get packed down into cracks and crevasses below the surface? The Soviet observers think that the surface is fairly strong, but it is my opinion that there is nothing in the pictures that is very reassuring. However, at least a solid object landed on the lunar surface and did not sink in too deeply.

Luna X orbited about the Moon and at the nearest point to the Moon it was approximately 350 kilometers above the surface, and at the farthest point about 1,000 kilometers from the surface. This orbiting vehicle had a gamma-ray counter on it which was able to record gamma rays of different wavelengths, particularly those of potassium, uranium, and thorium. Soviet observers maintain that this counter shows that the surface of the Moon is somewhat like terrestrial basalts. Terrestrial basalts are those that are produced in lava flows on the surface of the Earth, and this would seem to indicate that the surface of the Moon did originate from lava flows. If this is true, both with respect to the maria and the mountainous areas, the Moon's surface has been highly differentiated like that of the Earth; and of course, if this is true, like that of the Earth, the very early history of the solar system is hidden from us by lava flows. This will be most disappointing if it is the case, for we will have lost the early record of the lunar history as we have lost the early record of the Earth's history. However, preliminary reports indicate that the potassium content is on the low side of basalts and possibly extensive lava flows do not exist. Uranium and thorium were not detected, and reports indicate that the maria and the mountainous areas have the same composition. It is difficult to interpret these reports, and until the detailed studies from the U.S.S.R. are published we will not be able to answer these questions with confidence.

On June 1, 1966, Surveyor 1 made a soft landing on the Moon and sent back, during the next two weeks, some thousands of pictures of the lunar surface. This was a remarkable engineering accomplishment, which gave us our best detailed pictures of the Moon. These showed that the Moon is covered with some sort of rubble of rather low physical strength. This may consist of very fine dust, a sandy type of material, or such material mixed with larger rocks. This is indicated by a well-formed crater of moderate size, a few meters in diameter, with a sharp rim. The supporting pads of this spacecraft sank slightly into the lunar surface, indicating rather low physical strength, and they threw up material which is of darker hue than the undisturbed surface in the immediate neighborhood. This was contrary to expectation, for we supposed that the surface was blackened with respect to the material below and not the reverse. This suggests that probably the material in the neighborhood of the landing site of Surveyor 1, namely, the walled plain near Flamsteed, may contain carbonaceous material similar to that observed in the carbonaceous meteorites. Objects of most curious shape are scattered about the surface, some of a rather sharp angular character and some very rounded in appearance. Suggestions in regard to the chemical composition and physical strength of these vary widely, all the way from rather insubstantial rocky material or pumice to iron-nickel. Of course, no one can determine uniquely the chemical composition of strange objects and strange surface material in a strange environment by looking at pictures.

Possibly we must wait until the Apollo landings are made and we have actual samples of the lunar surface before we are able to unravel the Moon's early history. It may be that, if the Moon is highly differentiated, it will be necessary to go to Mars or Venus in order to learn more about the early history of the solar system. This, of course, will require many years and will not be done quickly at all.

Men from the beginning of history have marveled at what the Moon is and how it originated and how the solar system originated. It is my expectation that our space exploration will do much to explain what this past history has been, and I shall be fascinated by the study. I believe that both scientists and laymen will be much interested in the results. All of us wish we could know the future. This is impossible, and the next most interesting time is the past, both the history of men on Earth and the long and fascinating history of the Earth, planets, the solar system, and the stars.

15

The Lunar Surface

GERARD P. KUIPER

Study of the lunar surface dates back to the year 1610, when Galileo turned his newly made telescope toward the Moon. With a series of elegant and masterful arguments, he swept away the ancient misconceptions, demonstrating that the Moon possessed a very rough and mountainous surface. He showed that the darker lunar areas are comparatively smooth and frequently bordered by lofty mountain ranges and that the brighter areas contain numerous circular depressions of all sizes, each bordered by a ring of mountains.

Subsequent telescopic investigations were mainly concerned with mapping the Moon and providing an acceptable system of nomenclature, although early in this period Hooke emphasized the similarity between lunar craters and circular features left, first, by bursting bubbles on the surface of boiling alabaster and, second, by small, heavy objects being dropped into a pipe-clay mixture. He favored the first explanation for the lunar craters, assuming some type of volcanic exhalation to be analogous to the rising and bursting of bubbles.

After a hundred years of little progress in lunar studies, the beginning of the nineteenth century saw a sudden resurgence of interest in the Moon, resulting in much more accurate and detailed maps and many measurements of the heights, depths, and positions of lunar features. In addition, the foundations for the oft-mentioned volcanic-versus-impact theories for the formation of the lunar craters were laid at this time.

The last decade has seen a tremendous increase in lunar investigation. This has been largely due to the advent of the space age, the Moon being the nearest and most accessible space target. Not only have Earth-based investigations been stepped up, but circumnavigation and hard and soft landings by unmanned spacecraft have now been successfully accomplished.

Despite these large gains in data acquisition, our understanding of the Moon is far from complete. Nevertheless, each new finding adds to the overall picture, and the result has been the increased recognition that both volcanic processes and meteoritic impacts have left their marks. Volcanism does not explain the telescopic craters, as had been assumed, but has acted to fill the mare basins and has left numerous smaller structures both on the maria and on the terrae that are unmistakably of igneous origin. Meteoritic impact appears to be the cause of the large circular mare basins (which filled with lavas later), their associated mountain chains, the great majority, if not all, of the telescopic craters, and the sharply defined small craters.

The bright surroundings of the maria—the *terrae*—are found to be higher than the maria, some 1–3 kilometers above them. The terra elevations scatter widely, reaching several kilometers in the mountain ranges. The most prominent of these ranges occur as peripheral mountain chains around the near-circular maria. Examples are the Apennines, the Alps, the Carpathians, and the Altai Scarp. These arcuate chains surround the maria as the crater walls surround crater floors, an analogy that can be carried further and implies, apart from scale, a similar origin. This origin is probably impact by massive objects. In the case of the impact maria and pre-mare craters, the source of the objects may have been a satellite ring around the Earth through which the Moon swept very early in its history, in its outward journey from its position of origin very near the Earth. The post-mare craters, on the other hand, are presumably mostly asteroidal in origin, as is the case for the craters observed by Mariner IV on Mars. The relative crater numbers, Moon versus Mars, agree with this explanation. The crater density on Mars is about fifteen times that on the lunar maria, resulting from the closeness of Mars to the asteroid ring.

While the circular symmetry and the arcuate walls surrounding the maria betray their origin as due to gigantic impacts, the subsequent flooding of the impact basins calls for a separate explanation. The presence of smaller impact craters on the inner slopes of the bowl-like basins—craters that were later damaged by invading lavas—shows that the filling of the basins was not immediate but often after a considerable lapse of time, in some cases perhaps as long as a million years. The lavas are therefore assumed to have been generated internally, by radioactivity, the same heat source that was responsible for the melting of the asteroids, the parent bodies of the meteorites. That lava streams indeed covered the maria is seen from the numerous prominent flows observed on Mare Imbrium and Mare Serenitatis.

These lunar lava flows are very extensive. Several are from 50 to 100 kilometers long; one flow on Mare Imbrium approaches 200 kilometers. They are bounded by fairly steep flow fronts that stand out when observed with low sun angles, either as shadow bands or brightly illuminated strips. The thickness of these large flows varies from about 50 meters to well over 100 meters. They each have a characteristic color, neighboring flows often having different colors. Recently, at the Catalina Observatory, it was found that the flows near the center of Mare Imbrium originate from a row of small volcanoes situated along the main ridge crossing the mare. Apparently the mare ridges, related to the geometry of the mare basins either as radial or concentric structures, were the sources of the last phases of lava deposition. The volumes of these last deposits are enormous.

Lavas deposited in a vacuum and at the low surface gravity of the Moon will be highly vesicular, with a bulk density of 0.1–0.3 gram per cubic centimeter at the surface and slowly increasing inward as the hydrostatic pressure increases. Meteoritic particles of one gram or less impacting on such cellular material will bury themselves, but large masses penetrating into the deeper solid rock will produce open craters. Fragments ejected from large impact craters will bury themselves upon striking the Moon if they are small and arrive from great distances; large ejecta will produce secondary craters. The blankets of ejecta around major impact craters will therefore not extend indefinitely, thinning out with distance, but will have reasonably well-defined outer boundaries at which the surface density of the blanket can no longer be "absorbed" by the highly porous surface lavas. The expected result—the ejecta blankets around major impact craters having rather well-defined outer boundaries—has been confirmed by observation.

In our laboratories it is possible to measure, weigh, and chemically analyze substances and to determine their bearing strength, crystal structure, heat conductivity, and numerous other physical properties. By contrast, nearly all our information about the Moon is based on photographic images, taken from the Earth by means of powerful telescopes, from spacecraft in flight, or by soft-landers that took pictures of the surroundings afterward. If the Moon were a world totally unrelated to the Earth, the pictures obtained might be very puzzling. The fact is that its surface features very closely resemble the mountains, craters, lava flows, and numerous substructures seen on Earth. There are some differences, but this should cause no surprise. The Earth, unlike the Moon, has a substantial atmosphere and one which contains water vapor and rain; wind and water are powerful agents modifying the terrestrial landscape by erosion. But other causes of erosion are common to both planets: impacts by large and smaller meteorites, volcanism, and various processes of mountain formation. It is therefore necessary to compare the lunar photographs with terrestrial surface structures that have not yet been subjected to erosional forces that must be absent on the Moon. This means that very recent volcanoes, very recent lava flows, and very recent meteorite craters on the Earth will contain clues important to the understanding of similar structures on the lunar surface. The problems of interpreting the lunar photographs are therefore problems of "photo interpretation," with suitable photographs of recent geological structures being of paramount importance.

There is another point to be noted. Because the overwhelming majority of telescopic craters are almost certainly due to impacts, the areas on the Moon showing the densest crater cover must be the oldest and the areas showing the lightest cover, the youngest. Now the dark areas, or maria, systematically show a very much lower crater density than the lighter-colored highlands or terrae. It follows that the maria are younger than the terrae. The role of the lunar scientist resembles the role of an archeologist who studies a very ancient site of human habitation. He peels off each cultural layer in turn and so attempts to reconstruct the human activity of each era. The student of the Moon must do likewise. He starts mapping and interpreting the maria and then proceeds to older and older structures on the terrae. The numerous smaller impacts that have occurred everywhere supply, at least quantitatively, a time sequence. The calibration of this time sequence in terms of years or of terrestrial events must be done in the wider context of planetary and asteroidal studies.

Returning now to the nature of the mare surface: I have already noted one important difference between Earth and Moon. Lunar lavas will be extremely porous at the surface, like rock froth. Other differences will be caused by greatly reduced surface gravity on the Moon, only one sixth of that on the Earth. This will mean that ejecta from impact craters can be thrown to greater distances. Also, because of the absence of a lunar atmosphere, small particles will stay with the large particles in their trajectories and not be slowed down differentially. Finally, lateral drifts caused by terrestrial winds carrying small ejected particles, such as are observed around Meteor Crater in Arizona (where a southwest wind caused an accumulation northeast of the crater), will be absent on the Moon.

The three successful Ranger flights extended the optical resolving power with which sample areas of the Moon can be studied about a thousand fold, from about 0.5 kilometer for the best telescopic photographs to about 0.5 meter for the last of each of the three Ranger missions. Thus, they have closed the worst information gap about the lunar surface topography. In 1966 the successful Luna IX and Surveyor 1 missions extended the optical resolution to about one millimeter.

The bewildering amount of surface detail shown in the lunar photography—Earth-based, Ranger-based, and from the surface landers—makes it necessary to look first to those features that have a diagnostic significance. A lava flow is such a structure. Its length, width, thickness, mean slope, and surface texture are all diagnostic of a particular type of volcanic process. A well-defined meteorite crater, not marred by subsequent events, is another structure of known and distinct origin. A graben, being a narrow strip of terrain bounded by two vertical walls along which the strip appears to have sunken by 50, 100, or perhaps 200 meters, is another type of structure of unmistakable identity. The process of lunar exploration therefore proceeds by identifying all structures whose origins are basically clear from terrestrial analogues. This process, followed through systematically, appears to account for nearly all structural elements observed on the maria. The highlands, however, with their longer and very complicated history, which appears to have involved local subsurface melting and partial subsidence, still presents many open questions. In the following paragraphs I shall enumerate those structural elements of the maria whose origins and modes of formation appear reasonably clear.

First, the ray craters and the associated *crater rays.* The Ranger VII records are especially useful here because they cover an area traversed by two systems of crater rays, one belonging with the crater Tycho and the other with Copernicus. Earth-based photography had already disclosed that major crater rays tend to be broken up into *ray elements,* each often a dozen kilometers or so in length, which together produce the ray pattern. The Ranger records disclosed that at the head of these ray elements there normally exists a single white crater or, more frequently, a small cluster of white craters, from which the ray element issues in a direction away from the central crater, be it Tycho or Copernicus. Analyzing the possible explanations for these phenomena, I believe that Tycho and Copernicus both resulted from *cometary impacts,* with the ray elements caused by the impacts of cometary debris. This debris produced the small white craters or clusters of craters, and the main blast from the central explosion caused by impact of the comet nucleus drove away the ejecta from the associated impacts in directions radiating away from the central explosion.

Ranger VIII covered an area near the southern shore of Mare Tranquillitatis. This mare is traversed by weak rays issuing from the giant crater Theophilus. Analysis of the Theophilus rays on Mare Tranquillitatis has shown that they differ from the more prominent ray systems of Tycho and Copernicus. Ray elements are not well developed and do not have white craters at the head. Instead, the rays, apart from giving slight local increases of reflectivity, do not appear to disturb the lunar surface in the least. Thus I believe that Theophilus was not due to a cometary impact that was accompanied by a large cloud of secondaries, but to a single body, either asteroidal or cometary, that merely sprayed a thin cover of light material in a ray-like pattern over its surroundings from the central explosion.

An examination of the properties of the comet family in the solar system has disclosed that the probability of impacts by periodic comets is small compared to the probability of impacts by new, parabolic comets penetrating the interior of the planetary system for the first time. Because these parabolic comets are likely to have high relative velocities with respect to the Moon (around 50 kilometers a second), the impacts, per unit mass, will be violent. The average mass of a typical parabolic comet has been estimated to be 1018 grams.

The Ranger VII photographs also disclosed the presence of *white mountains* on the Moon, composed of sharp-crested, near-linear ridges, usually oriented along the structural pattern of the mare (defined by the much lower wrinkle ridges) and on whose crests are seen several small white craters. It is assumed that these white mountains are volcanic in nature and that during the final stages of formation they were covered with sublimate. Such sublimates also form on terrestrial volcanic ridges where they are normally destroyed promptly by solution in rain water. An example of a white mountain on the Earth is Laimana Crater, Hawaii. Here sublimates consist of sulfates, sulfides, chlorides, carbonates, and oxides, among others. On the Moon there will be no losses by solution in rain water but the lunar vacuum will reduce the deposit by evaporation so that only substances with very low vapor pressures at the maximum temperature (100 degrees Celsius) will survive.

I have already remarked that the mare ridges (or wrinkle ridges) are structually related to the geometry of the mare basins in which they occur. On the Ranger VII records, one impact crater may be observed with fair resolution formed squarely on one of the ridges in Mare Cognitum. From these records we have concluded that the ridges are due to dikes extruded into fissures caused by both global and mare-wide tensional forces, with these dikes branching and also spreading locally as sills or laccoliths, thus causing *en echelon*, or braided ridge, structures and also giving the ridges locally considerable width. These broad expanses do not appear to contain depressions or dimple craters, consistent with the explanation of these features as due to collapse.

Collapse depressions in terrestrial lava fields result from the withdrawal of subsurface lavas under a solidified crust during the last phases of lava deposition. The actual presence of subsurface caves or tunnels in the lunar maria is indicated by the central sinks or caves seen in some dimple craters. The interest in lunar caves may develop in preparation for manned landings, providing natural shelters against solar plasma ejected from active solar regions. The coverage of maria and lava lakes with collapse depressions, while common, is not universal as shown by their absence on the strip of Mare Nubium observed by Ranger IX.

Collapse depressions on the Earth are most numerous near flow fronts where breaks in the walls can occur through which the remaining fluid lavas drain and form secondary flow units. No clear flow fronts are observed on many of the Ranger pictures. Some of the lunar collapse depressions may therefore be caused by the release of water vapor and other gases rather than by the drainage of subsurface lavas.

The bearing strength of the mare surface is of obvious significance to the landing of manned spacecraft. Terrestrial laboratory samples of rock froth produced in a vacuum have a bearing strength roughly 3 kilograms per square centimeter (3 tons per square foot). Natural samples of rock froth (reticulite), solidified in free fall from liquid basaltic lava ejected by the Hawaiian crater Laimana in 1960, have an almost identical limiting bearing strength. In the evaluation of the Ranger IX records some fifty rocks were found near a primary impact crater from which they were apparently ejected. The crater is 50 meters in diameter; the rocks are seen up to distances of 120 meters. They average about 1 meter in diameter and are mostly buried in the lunar surface layers, projecting only some 20 centimeters above the surface (as determined from shadow measurements). From these data the limiting bearing strength for the upper 50 centimeters of the floor of Alphonsus is found to be about 1–2 kilograms per square centimeter. This recurrent figure is about a thousand times smaller than the limiting bearing strength of solid basalt. It resembles the bearing strength of wet beach sand. It is adequate for landing operations.

The successful landings of Luna IX and Surveyor 1 have confirmed the suitability of the lunar mare surface for landing operations. In the case of Surveyor 1, a special effort has been made to measure the bearing strength with precision. It was found to be between 0.4 and 0.7 kilogram per square centimeter, averaged for the upper 30 centimeters, somewhat smaller than but of the same order of magnitude as the figure found for the Alphonsus floor for the upper 50 centimeters.

I would, in concluding this section, summarize the Moon's surface features in the following way. The surface structures are all unmistakably indicators of the lunar maria as igneous deposits. The similarities of the frequencies of collapse depressions and dimple craters in the three Ranger impact areas has come as a surprise. It should not be inferred, however, that this necessarily implies that every mare region has nearly identical properties. In fact, a portion of Mare Nubium that has been well observed by Ranger IX does not possess collapse depressions though numerous lava lakes in a widely scattered area near the center of the lunar disk all possess such depressions. The presence of dark-halo craters in certain districts of several maria and their identification as explosive maars also indicates a non-uniformity of surface properties even within the maria. Such differences should cause no surprise if compared with the very rich variety of volcanic phenomena on terrestrial lava fields. The conclusion is obvious: a much more extensive study of lunar surface detail will be required before the lunar maria can be considered reasonably known.

Now I should like to turn to the results of the two successful soft landings on the Moon accomplished by Luna IX in February, 1966, and by Surveyor 1 in June, 1966. Luna IX landed on the edge of Oceanus Procellarum, among the small hills southeast of Cavelerius F. Three complete 360-degree horizontal scans were obtained, with the Sun 7, 14, and 27 degrees above the horizon, and part of a fourth view with 41-degree sun elevation. The camera opening was approximately 60 centimeters above the lunar surface with the axis tilted about 20 degrees with respect to the vertical. As a result, an arc of the lunar surface somewhat in excess of 180 degrees was observed.

The resolution in the foreground of the Luna IX panoramas is extraordinarily good, roughly 2 millimeters. The texture of the lunar surface is remarkably homogeneous and everywhere that of nearly level but highly vesicular lava, with bubble sizes below about 1 centimeter. A number of small craters and depressions occur in the panoramic view. The craters that appear to be caused by impacts have rough floors of the same texture as the surrounding lava surface and are reminiscent of artificial terrestrial craters in basalt. The depressions are dimple-shaped. A number of profiles have been measured by Russian scientists. Several sharp riblike ridges are seen which almost certainly were formed by magmatic intrusions into cracks during the last phases of the mare deposition.

Surveyor 1 landed within the roughly circular ghost crater on whose southern edge is found the crater Flamsteed. The Surveyor 1 field differs appreciably in appearance from that of Luna IX. While numerous delicate lineaments are visible, signifying the presence of partly visible igneous deposits, there is an overlying blanket of different material, locally perhaps 30 to 60 centimeters thick in which are embedded an appreciable number of rocks. A close view of one of these rocks, some 50 centimeters long, with a resolution approaching 1 millimeter, indicates that it is not a fragment but may be a volcanic bomb. Another part of the horizon is covered with numerous angular blocks, apparently fragments. It appears that Surveyor 1 landed just outside the west outer slope of an impact crater provisionally identified as a small white spot on Earth-based full-moon photographs. The crater itself may be around 100 meters in diameter. The fine material in which the rocks east of the spacecraft are seen embedded must, in part, be debris from this crater but, because of the rather uniformly fine texture near the spacecraft, an admixture of volcanic ash may be present as well. The apparent infrequency of collapse depressions (the absence is a property of volcanic ash fields) in the immediate vicinity of the spacecraft would support the ash hypothesis.

The limiting bearing strength of the lunar surface, as determined from the comparatively shallow imprints of the supporting pads of the spacecraft into the Moon, is slightly smaller but of the same order of magnitude as that found from the Ranger IX data. In other words, the general predictions on the absence of loose dust and on the bearing strength of the lunar surface made by the writer have been confirmed by the soft landings.

With inclusion of the three Ranger impact areas, five mare-type regions of the Moon are now known with resolutions below 1 meter. All five regions give evidence of an igneous layer either at the visible surface or else just below it with a blanket whose thickness may vary from a few inches to perhaps a meter. From the limited data at hand, it is not possible to say what fraction of the irregular, thin blanket is due to displaced material from primary impact craters and what fraction due to volcanic ash deposited during the terminal phase of mare formation. From the integrated volume of the primary craters, the average crater debris deposit is found to be well below 1 meter, with most of it expected to be concentrated within one or two crater radii from the crater themselves and part of the remainder buried below the visible lunar surface. A much larger sample of the lunar surface will have to be known with 1-meter resolution or better before the relative importance of ash deposits and crater debris can be ascertained, and before the surface distribution of collapse depressions is known.

From a scientific point of view, therefore, the acquisition of greatly expanded coverage of the lunar surface with resolutions of about 1 meter appears to be a first prerequisite. Such a survey was the objective of the Orbiter program. It was designed to give a broad representative coverage of the equatorial band of the visible hemisphere. It cuts across maria and terrae, contains the Ranger VIII and IX impact areas, and will strengthen the empirical basis from which to interpret the history of the lunar surface. Perhaps next in priority is a much more detailed knowledge of the reverse side of the Moon. The Luna III coverage was made under full-moon illumination, so that no relief could be observed, while the resolution was limited to about 10 kilometers. While the results of Luna III were most impressive at the time (1959), they do not fully solve the important problem of the distribution of the maria on the reverse side. The Zond III results are much more detailed but relate only to a limited sector. They suffice to show, however, by the extraordinary arrangement of crater chains observed, that coverage of the entire lunar surface with at least 1 kilometer resolution is essential.

A detailed physical exploration of the lunar surface will have to await manned landings and the installation of delicate recording and measuring equipment on the lunar surface. Such data will be needed in answering the broader questions of lunar constitution, history, and relationship to the Earth.

III

OTHER PLANETS AND OBJECTS

16

Mars

DONALD U. WISE

The study of planetology is nearly as old as science itself in that much of man's early scientific efforts involved attempts to understand his own planet. We have come a long way in the study of planet Earth, of the forces that shape it and of the ways in which we can control and use them. One obvious way to increase this understanding is through comparative planetology, the comparison of details in the development and history of the Earth with those of other planets.

The groundwork of comparative planetology was laid by patient observations of astronomers over many centuries. In the last few years the field has entered a new phase, that of physical astronomy. Spacecraft are at hand to carry instruments, and ultimately men, to the Moon and planets to begin physical exploration rather than merely to look at them from a great distance. As one of the first steps in this exploration, the historic space probe Mariner IV was launched to pass within 7,000 miles of the planet Mars in July, 1965. To make the rendezvous the craft had to travel a quarter of a billion miles during the previous seven months. During its half-hour encounter, Mariner IV took photographs and made a variety of measurements which significantly affected our knowledge of Mars.

Mars like Earth is one of the four, small, dense, terrestrial planets nestled in close to the Sun, Earth being number three in the sequence and Mars being number four. For Mars the day is twenty-five hours long, and its year is about twice the length of ours. Its axis is tilted about as our own, giving it alternating summer and winter seasons. In the winter hemisphere thin polar caps of ice or possibly of solid carbon dioxide spread out more than half way toward the equator, only to retreat and virtually disappear in the following summer. Several types of clouds are visible periodically on Mars, including great yellowish dust clouds blowing rapidly across the surface, sometimes completely obscuring the surface.

Permanent dark areas and lines break the generally orangeish surface of the planet into many smaller sub-areas. In the Martian springtime these darker areas turn progressively green-gray and spread outward from the polar regions toward the equator. With summer and autumn they turn brownish and have lower contrast with the orange areas. This pattern of color change, suggestive of the behavior of vegetation, is a major reason for speculation that some form of life might exist on Mars.

Many significant differences exist between Mars and the Earth, one of the most far reaching being the absence of oceans. Water is scarce on Mars at best, and it is doubtful that any water occurs in liquid form. The atmospheric pressure is so low that only ice and vapor are stable. A second great difference, revealed by Mariner IV, is that Mars has a comparatively dead interior more like that of the Moon than the Earth. Thus, the understanding of Mars involves working out a model of planetary surface less actively churned from the interior than the Earth's surface and having no oceans or running water to sculpture it, redistribute material, or make chemical separations.

The Mariner IV photographs of the Martian surface showed no linear mountain systems as we know them on Earth, but rather a pock-marked and cratered surface with craters up to 185 kilometers in diameter. Some of these could have been formed by volcanic processes, but large numbers must have been formed by meteorite impacts. The position of Mars next to the asteroid belt, the source of most of the meteorites, must produce a greater number of impacts per unit area per unit time than on the Moon or Earth. Martian craters have central peaks in the same proportion as do lunar craters of apparent impact origin. The dominance of these externally produced features, produced slowly over geologic time, places limits on the relative rates of erosion and change on the Martian surface. They stand in sharp contrast to the Earth, where the continuous plowing of the surface by mountain-building and erosion has destroyed or camouflaged all but the most recent craters.

The Martian craters show some changes from the young lunar impact craters in the smoothing of rough topography and a filling in of the floors. The most likely agent for these changes is the wind operating in the thin Martian atmosphere to erode and redistribute material. The periodic dust clouds moving over the surface indicate the presence of the process, but the continued existence of the craters reveals that it is comparatively ineffective.

Redistribution of surface material by wind action on Mars must be related to the circulation pattern of its atmosphere, a pattern which could be reflected in the orientation of sand-dune complexes provided that we could see them. In future higher-resolution photographs it will be desirable to check for the existence of dune complexes, with their clues to the directions of atmospheric and sediment movement. It would be a mistake, however, to think of dunes as the only or indeed the most likely form of wind-blown deposit. If sand continues to be agitated by the wind, smaller and smaller particles are produced. On Earth these particles are either blown away to settle elsewhere as thick, wind-blown dust deposits called "loess" or are incorporated into water-laid sediments. On Mars, with its slow rate of erosion, almost all the particles could be reduced to dust, producing great loess plains as a major surface feature.

The final collecting place for the wind-blown dust, if such a place exists on Mars, would be the regions toward which surface winds blow most strongly and frequently. Because the equatorial regions of a planet are more strongly heated by the Sun than the polar regions, these warmer areas are more likely to be characterized by rising air, and the polar regions by sinking air. Thus, surface winds would have a tendency to move material out of the higher latitudes. Collection of wind-blown sediments in equatorial regions has been one of many suggestions for the cause of a darker zone near the equator of Mars. Since this darker zone is offset somewhat toward the southern hemisphere, unequal wind patterns in the two hemispheres would be necessary to produce it. The highly elliptical orbit of Mars makes the temperatures and lengths of seasons quite unequal in the two hemispheres, permitting this kind of interpretation although the validity of the hypothesis is still questionable at best.

If there are locations for selective deposition on Mars, there must also be regions of selective removal of material. On the basis of our previous arguments about planetary surface winds, these places should be the middle and higher latitudes. Erosional regions, marked by selective removal of softer or weaker materials, would leave more resistant fragments sitting on protected pedestals of the softer materials. The most likely caps of these pedestals are meteorites, volcanic materials, or clumps of material thrown out by meteorite cratering. This contrast in small-scale landforms between relatively featureless plains versus pedestal areas may be an excellent clue to the local wind behavior. In general we would expect the smooth plains to be more common in the low latitudes and the pedestals to occur in the middle latitudes.

The pattern of erosion and deposition on Mars would not be limited in elevation in the same way as on Earth. On our planet, wind erosion can go no deeper than sea level except in a few desert basins, which are maintained below sea level by excessive evaporation. On Mars this limiting, sea-level base of erosion could not exist. Instead, erosion would cut deeper and deeper until the depression was sheltered from the wind and erosion would cease. Alternatively, the planet could become so misshapen that plastic flow of its interior or volcanic activity would begin to restore it to its proper ellipsoidal shape.

Our best measurements of the shape of Mars suggest that it bulges around the equator approximately twice as much as expected for a planet of that size and rotation rate. Some redistribution of material toward the equator by past wind action could be a contributing factor to this bulge. However, the measurements of flattening are so difficult and open to question that at present it would be well to withhold judgment on this.

As actively operating surface processes on Mars would have destroyed most of the craters, so would an actively moving interior. These internal movements, driven largely by internal heat of the planet, would have wrinkled its surface into linear mountain systems, would have slipped one portion of the crust past another to produce major offsets of craters, or would have built vast volcanic fields and cones. The seeming lack of these features on the Mariner photographs does not say that internal processes are inoperative on Mars—only that they are not the dominant processes. When meteorite-produced craters can be separated from volcanic ones, the exact level of internal activity can be assessed more clearly. The landing of a recording seismograph on some future mission will provide some of the best answers to this question.

Supporting evidence for a relatively dead interior of Mars was derived from the magnetometers carried aboard Mariner IV. The magnetometers never recorded any change from the interplanetary magnetic background even in passing close to the planet. This places an upper limit on the strength of any magnetic field of Mars as being about one thousandth of that of the Earth. Because one of the requirements for a magnetic field seems to be internal motion of an electrically conducting core, this lack of a magnetic field also implies a relatively quiescent interior for Mars.

The fact that the interior of Mars is comparatively dead in contrast to that of Earth does not imply a complete lack of thermal energy there. The thermal energy of a planetary interior comes largely from the decay of radioactive materials. Thus, the heat produced by a planetary body depends on its volume while the heat flow through its surface depends on its surface area. It follows that the heat passing out through an area, say a square meter of surface, varies as the radius of the planet. Accordingly, if radioactive concentrations were the same and thermal equilibrium was obtained, Mars should have twice the rate of heat flow as the Moon and half the heat flow of the Earth. Evidence on the Moon of the escape of internal energy is abundant in the form of lines of volcanic craters along fracture zones and by the apparent lava fields. If Mars has radioactive concentration similar to the Earth or the Moon, it would be surprising if we did not find features revealing the escape of this internal energy, developed at least as extensively as on the Moon but not nearly so prominently as on the Earth.

Size alone is an additional factor which might make Mars seem comparatively inactive in contrast to the larger Earth. The greater the distance that heat must move by conduction from the interior to the surface of a planet, the more difficult conduction becomes. Instead, the heat will be transferred by actual movement or flow of the material circulating between different levels of the interior. Even in the extreme case—i.e., that Mars had just as much heat flowing through a unit of surface area as Earth—the smaller Mars might be able to dissipate the heat by conduction rather than by an actual churning of its interior. Thus, it would appear to have a less active interior than the Earth.

Using this frame of reference of Mars as a body of thermal activity intermediate in intensity between that of the Moon and Earth, a consistent explanation can be developed for the systems of fine lines, sometimes called canals, which crisscross the Martian surface. The lines would be deep-seated fractures of the planet's crust, still active but possibly formed a very long time ago. Once formed they could continue in existence almost indefinitely in the absence of any churning or disruption of the deeper parts of the planet beneath them. They would become permanent channelways for the escape of volcanic gases, lava, water, and of heat from the interior. The escaping materials would darken the surface along the line, probably pitting it with small vents, but not necessarily building any great volcanoes along it. If larger volcanic features exist, they are more likely to form in the larger channelways provided by the intersection of these fractures. If vegetation is present, it would be most likely to develop more densely along the zone as a result of the more abundant water and of the more intense shattering to provide better or deeper soil for rooting. The result would be further darkening and enhancement of the visibility of the zone.

Lines of small volcanic vents are abundant on the Moon, but are much less prominent in their development. If Mars with its greater size has a correspondingly greater heat loss per unit area, similar lines of craters would be expected to be more strongly developed in order to pass off the extra heat, and they would correspondingly stand out more clearly. In this model, the Earth would show extreme development of lines of vents were it not for its shifting interior, which causes the deep-seated sources to abandon the fracture lines, and these lines are then slowly destroyed by erosion or mountain-building.

The Mariner IV photographs were incapable of resolving small-scale thermal features, but we will certainly want to examine future photographs for them. It would also be well to recall that we were a long time in recognizing these features on the Moon. Future Mars missions will certainly run infrared thermal scans to determine whether these lines show up as very hot zones. Far in the future, after landings become possible, we will want to make direct measurement of the rate at which heat is flowing through the Martian surface. This, along with seismic measurements, is the key to understanding the internal energy balance of the planet, to learning just how inactive it actually is, to comparing its radioactive content with that of the Earth and the Moon, and thus to finding out whether they came out of similar or different chemical crucibles in the origin of the solar system. For the present, however, we must be content with the search for areas of thermal escape from the Martian interior.

The absence of oceans on Mars has a much more profound effect on the nature of the planet than merely making wind the dominant surface agent. On Earth the oceans have provided the site for major chemical differentiation of portions of the crust. Consider that areas of beach sands are nothing but pure silicon dioxide. In contrast, limestone areas and dolomites are nothing but magnesium and calcium carbonates. The list is long, but for the most part these sedimentary rocks represent highly selective concentration by ocean waters of certain materials originally derived from volcanic rocks and from the atmosphere. In that these marine sediments are reworked into granite-cored mountain ranges which become part of the great granite rafts which we call continents, it may be that one of the necessary conditions for the formation of continents is the existence of oceans in which the extreme chemical differentiation of materials can take place.

It is not enough for the chemical changes to take place at the surface. There must also be a method of supplying fresh material and removing the chemically altered material by burial. On Earth this process operates quite well through the steady plowing of the surface by erosion and mountain-building. On Mars some changes could certainly occur in the wind-blown dust as it moves through the atmosphere. However, such changes would probably be relatively minor and, in the absence of water, much less efficient than the Earth's chemical differentiation of sediments by oceans. Further impediment to change derives from the lack of an efficient method of burial of the chemically altered sediments, once they are formed on Mars, or of the supply of fresh material by the over-all plowing of the planet's surface. The surface dust could become thoroughly oxidized as some of the polarimetric determinations of Mars have suggested, but there would be no efficient way of removing vast quantities of materials from the surface. Thus it is much more likely that a small area of the surface of Mars would contain more representative samples of the average chemical composition of the over-all Martian surface than a similar random area of the Earth's surface could represent the Earth's average surface composition.

The absence of oceans on Mars also has major effects on the composition of its atmosphere. The atmosphere and hydrosphere of a planet represent the dregs or remnants of any primordial atmosphere plus all the volcanic gases which continually escape from the interior, plus any gases released by rock weathering, less any gases which escape to space or are chemically combined and buried in sediments. On Earth, if we consider the amount of former atmospheric carbon dioxide that is now tied up in limestones, it appears that the atmosphere would be dominated by carbon dioxide were it not for the ocean's ability to remove it as limestone.

The evolution of the Martian atmosphere differs from the Earth's in the relative efficiency of the chemical combination process versus the escape to space. Comparatively little of the volcanic gas escapes to space from the Earth; the bulk goes into the oceans and sediments. Mars's volcanic gases may well be escaping from its interior at a slower rate than the Earth's because of the less active interior. A bit of this gaseous material would remain behind as the polar caps, in permafrost or combined in sparse sediments, but a greater percentage than on Earth would escape to space because of the lower escape velocity on Mars.

We know a few facts about the Martian atmosphere. It contains some water and carbon dioxide. Precisely how much is still uncertain, although the total of all constituents is about 1/200 of the Earth's atmospheric density. The atmospheric composition could include some radioactively generated argon, nitrogen, or any of a wide variety of other gases. However, it would be a mistake to argue by analogy that because the Earth's atmosphere contains mostly nitrogen the Martian atmosphere also should be largely nitrogen. Nitrogen is abundant in the Earth's atmosphere only because it is highly soluble and unlike carbon dioxide is not easily removed from the oceans as sediments. Thus, for a planet without this limestone-removal system, it is much more likely that carbon dioxide would exceed nitrogen and, if cold enough, the planet would ultimately collect the carbon dioxide as dry-ice polar caps. Recent theoretical studies lead to such a picture on the basis of thermal model calculations.

This discussion of the chemical nature of the Martian surface has been generalized simply because very few hard facts are at hand. Its purpose is to point out what a complex interacting chemical system the surface of a planet and its atmosphere can be. It is therefore important to determine the composition of the Martian atmosphere not only because of its significance for any life that might be there, but also because it constitutes such a far-reaching tool in working out the cumulative chemical effects of the surface over geologic time, including the type of volcanic emanations, the amount of chemical recombination going on at its surface, and the extent of escape of gases to space.

Of value and interest comparable to the exploration of space is the exploration of time. The exploration of time on our own planet has been in progress through geological researches for the past century and a half. We now know that the Earth and planets are four and a half billion years old. We know a good deal about the last eighth of the Earth's history and a few facts about the last three-quarters of its history. However, the first quarter is almost a complete mystery, even to the mere existence of rocks of that age. The continuing plowing of the surface has long since erased that ancient record. For the Moon, a much more complete record of the early events might still be diciphered. Unfortunately, the Moon is not a planet and may not have gone through the same complete sequence of events. For Mars, with its lesser activity and erosion, there is at least hope of finding evidence of its earlier history.

The key to working out the history of a body, be it Earth, Moon, or Mars, is the development of a stratigraphy, or rock sequence, and the order in which it was piled from oldest to youngest. On Earth we can use fossils and radioactive age dates to determine the sequence. On the Moon, techniques of interpreting which layer of rock overlaps which can be used to tell relative age based on the principle that an older rock unit must be present before another can be deposited over it. For Mars we are totally ignorant of any time sequence because the resolution of our view has been incapable of discriminating between various surface rock units. The Mariner photographs are at the barest limit of these geological capabilities. The obtaining of good photographs and other types of remote imagery of the entire surface of the planet is a key objective if we are even to plan efficiently for the study of the history of Mars.

The development of some kind of stratigraphy for Mars could provide clues to radical changes and significant landmarks in the history of the planet. Geologists refer to these as non-uniformitarian events or happenings for which the present conditions are not a clue to the past.

Evidence of past oceans is the most significant thing to seek among Martian non-uniformitarian events. I have already discussed the fact that chemical segregation of different rock units on Mars would not be particularly effective under present dry conditions. If, however, we should find ancient rock sequences with strong layering, including beds rich in carbonates or salt, the indications of past oceans would be quite clear. If there turns out to be life in some form on the planet, these data would have particular significance in telling us how this life originated and evolved. From a negative point of view, the evolution of life in the absence of oceanic waters would be equally interesting as a clue toward the evolutionary process. The probability of finding fossils in ancient sediments on Mars by using remote methods is so low as to make such efforts pointless. On the other hand, the existence of organic rich sediments in a layered sequence might be detected through remote methods, and these clues to the early evolutionary process would warrant further investigation.

Another non-uniformitarian event for which evidence might be sought is the existence of a thermal maximum on Mars. We now think that the planets had a relatively cool origin. Once formed, however, radioactive heat began to build up. Ultimately the most intense radioactivity decayed away, and the planet began to cool again. In the interim the planet went through a period of maximum internal temperatures, a thermal maximum. On the Moon, the apparent great lava flows which now form the maria may represent such an era of greater thermal activity. For Mars one would want to look at any stratigraphic column for progressive changes in composition of volcanic rocks as a function of time or for rock sequences characterized by a period of excessive lava outpourings on a planet-wide scale. If this turns out to be a general feature of planets, it may be part of the clue to the missing first quarter of the Earth's stratigraphic history.

What caused the numerous glacial eras in Earth's history is an unanswered question. In the history of Mars, times of greater surface water might have produced glacial polar caps. The clue would be ancient glacial deposits in the Martian record. If these are found, provided they ever existed, we could then contrast their times of formation with the age of glacial eras on Earth. In this way there could be a check on whether the causes of the ice ages were of a broad enough scale to have affected several parts of the solar system or were restricted to the Earth.

If the Martian canals are indeed fracture patterns, their orientations should bear some relationship to the rotational axis of the planet either present or past. On a slightly less active planet than the Earth it might be possible to determine how fracture patterns are related to the rotation by giving the system enough time to develop fully. This information would be a considerable aid in understanding how the Earth is operating today. From such evidence it might also be possible to find out when Mars tilted on its axis and to gain some clue to the mechanisms involved in this fundamental parameter of planetary behavior.

There is one other problem which I have mentioned only in passing because it is discussed more fully in another chapter: the search for life beyond the Earth. This is probably the most important single question to be asked on Mars. Does life exist there? What chemistry is it based on? What environment did it start in? What rates and mechanisms operated for its evolution? But the questions which I have been discussing, the physical nature and history of the planet, must be answered concurrently in order to know where to look for possible life and how that life fits into the environment and history of its planet. To look for life without the supporting planetary data would be like studying a garden without knowing of the existence of soil. The exploration of Mars must be a joint biological, geological, and geophysical effort.

The exploration of Mars has part of its reason purely and simply in man's curiosity and desire to know. It also has a practical side in contributing to our understanding of our own planet through comparative planetology. In effect, it helps to answer the seemingly impossible examination question, "Discuss the Earth and give two examples." Mars is our second example.

• 17

Venus

RICHARD M. GOLDSTEIN

The lovely light of Venus has engaged the imagination of men for millennia. As the orbit of Venus carries it from behind the Sun, it may be seen low in the western sky just after sunset. Each successive evening, after dusk, Venus will appear slightly higher in the sky, dim at first, but growing brighter nightly. After seven months the planet will reach its greatest angle from the Sun and then begin its westward motion back toward the Sun, still growing brighter. The time of greatest brilliancy will be one month later. Venus will then outshine all of the stars and other planets, being the brightest object in the sky, save the Sun and the Moon. Venus will be bright enough to cast a shadow at night and to be seen easily by the naked eye during the day if one knows where to look.

The westward motion of Venus will carry it into the solar glare in two more months, and it will be lost to the night sky for ten months; then Venus will repeat its eternal 575-day cycle. During its absence from the nighttime sky, Venus may be seen as the Morning Star in the eastern sky just before sunrise. There, Venus repeats the waxing and waning and the motion of the Evening Star, but in reverse order.

The ancients thought the Evening and Morning Stars were distinct entities, and the first was named Hesperus, the second Phosphoros. However, the Greeks of 500 B.C. knew they were alternate appearances of the same object, the discovery being attributed to Pythagoras.

Let me at this point summarize a few gross features of Venus.

Venus revolves around the Sun at about three quarters of the distance of the Earth, taking 225 days to complete each circuit. Thus, Venus and Earth return to the same position relative to the Sun every 575 days, accounting for the 575-day cycle of appearances in the morning and evening sky.

The diameter of Venus is estimated at 12,700 kilometers, compared to 13,200 kilometers for Earth. Venus is also slightly lighter than the Earth, having a relative density of 0.81. Because of these similarities, Venus has often been called Earth's sister planet. However, there are also pronounced differences. For example, the atmosphere of Venus is estimated to be ten to a hundred times more dense than the Earth's atmosphere. Another significant difference is the slow, backward rotation of Venus (one turn in 243 days). All of the planets revolve around the Sun in the same direction; all but Venus spin about their axes in this same direction also. The best of modern theories holds that the planets formed from a primordial, rotating mass of dust and gas, but it fails to account for the retrograde rotation of Venus.

Around 300 B.C. Heraclitus wrote that the planets, including the Earth, revolve about the Sun. This heliocentric view of the solar system was later abandoned by the Greeks, however, and for a very convincing reason. If the Earth did travel about the Sun, then the stars would be seen to shift their relative positions. Since no such shift was observed, one must conclude that the stars were millions of miles away—clearly an untenable conclusion. Much later, it became heresy to suggest that the Earth moves.

Copernicus restored the Sun, in theory, to the center of the solar system. He showed that the heliocentric assumption made it possible to predict with reasonable accuracy the positions of the planets. Because of the opposition to this possibility, Copernicus' work was published only after his death, in the middle of the sixteenth century.

The invention and application of the astronomical telescope by Galileo settled this question, and Venus provided the crucial evidence. For under the magnification of Galileo's telescope, Venus was resolved as a disk which went through the same phases as the Moon. Thus, the revolution of Venus about the Sun was demonstrated by the waxing of Venus from a thin crescent to a full disk, followed by the subsequent waning to the crescent.

Venus was destined to play another important role in man's

understanding of the solar system. By the seventeenth century, the shape of the solar system was fairly well known. Venus had been assigned position number two, farther from the Sun than Mercury, but not so far as the Earth. Observations of the apparent positions of the planets, coupled to the empirically derived laws of motion of Kepler, had led to good tables predicting their future angular positions.

Although the relative distances between the planets were known, the actual distances, or the scale of the solar system, were not. Thus an important astronomical constant was needed. This constant was defined as the mean Earth-Sun distance, and was named the "astronomical unit."

In 1639, a very young Englishman by the name of Jeremiah Horrox was busy calculating an ephemeris for Venus. His goal was twofold: to correct a small error in the tables of Kepler and to refute completely a rival set of tables. In accomplishing both of his goals, he discovered that Venus would transit across the disk of the Sun in 1639.

The transit of Venus is a very rare astronomical event, occurring in regular cycles of 243 years. The cycles are composed of four irregular intervals which alternate between short and long. The short ones are always 8 years apart, and the long ones are alternately 121.5 and 105.5 years. The eight-year interval is a consequence of the fact that the time for eight Earth years equals thirteen Venus years.

Horrox wrote of his discovery: "It induced me, in expectation of so grand a spectacle, to observe with increased attention. I pardon, in the meantime, the miserable arrogance of the Belgian astronomer, who has overloaded his useless tables with such unmerited praise . . . . deeming it a sufficient reward that I was thereby led to consider and foresee the appearance of Venus in the sun."

Thus Horrox was the first man to witness a transit of Venus. He used a telescope and brought an image of the Sun to sharp focus on a screen. By this experiment, he was able to calculate an improved ephemeris of Venus and to estimate that the astronomical unit is at least larger than 58 million miles—by far the largest estimate up to that time but still short of the modern value.

The next transits of Venus were in 1761 and 1769. By then, Newton had formulated his celebrated laws of motion and applied them to calculating the motions of the planets. The only remaining unknown was the scale factor. Thus, measuring the astronomical unit became the "final" problem of astronomy.

Halley had published a paper in 1716 describing how the forthcoming transit of Venus, if viewed from two separate and remote places on Earth, could be used to determine the astronomical unit to a high degree of precision. Because of the geometry of the situation, observers at different places on Earth would see Venus pass in front of different sectors of the Sun, and the astronomical unit could be calculated.

Halley never observed a transit of Venus, but his paper of 1716 created an avalanche of interest in the world of astronomy. Great expeditions were dispatched to remote areas of the Earth to observe the transits of 1761 and 1769. These expeditions were financed by governments, and thereby set the precedent of government involvement in large scientific projects; the occasion of these transits also set the precedent of rivalry in science between governments.

Altogether, over one hundred separate observations were made of these two transits of Venus. When the results were all in, the astronomical unit had not been found with the high accuracy which had been anticipated. The estimates ranged from 80 to 100 million miles. A major cause of the remaining uncertainty was the extensive atmosphere of Venus, which made it impossible to establish the exact moment when Venus entered upon the solar disk.

Venus has fulfilled her old promise, however, in the last few years. As we shall see subsequently, radar reflections from Venus have established a truly remarkably accurate value for the astronomical unit.

Venus is in many ways a planet of mystery. Many theories have been advanced to explain the scant observational evidence available, and they have been extremely disparate theories. The underlying cause for the fact that different, incompatible theories can be defended is this very lack of data.

The reason for these contradictory views is the dense cover of clouds which hides the surface of Venus from sight. The obscuring clouds themselves were the first certain feature observable with telescopes. By analogy with the clouds of Earth, astronomers at the turn of this century decided that the clouds must be made of water. And since no break in the clouds had ever been observed, they concluded that the clouds were very thick and that, consequently, the abundance of water on Venus was very great. The presence of so much water somehow suggested that Venus was a steamy swamp, perhaps populated as are our own rain forests.

To test this argument, spectroscopic studies were made to detect the characteristic signature of water vapor in the reflected sunlight. The early attempts failed to find any water vapor at all, and only the latest measurements show that, at least in the upper atmosphere, there is a very small amount of it.

But if the clouds are not water, what are they? An alternate idea arose that they are dust clouds, permanently stirred up from the surface by strong winds. Thus, Venus became a desert planet, dusty, windswept, and probably hot. This theory gained support in later years from the spectroscopic discovery of large amounts of carbon dioxide in the atmosphere of Venus. On our own planet, there is an equilibrium maintained between carbon dioxide in the atmosphere and carbonate minerals and quartz sand on the surface. However, this equilibrium is attained only in the presence of surface water. The large amount of carbon dioxide in the atmosphere of Venus was thus taken to indicate the absence of surface water.

A completely opposite point of view is possible and, in keeping with the divergent theories of Venus, was not overlooked. There is the possibility that no minerals are available because the surface is entirely covered by water. In this view, Venus is a featureless planet of oceans.

Still another explanation of the large amount of carbon dioxide in the atmosphere makes use of a presumed chemical balance at the time when Venus was formed. In the case of Earth, there was an excess of water over hydrocarbons, leading to our atmosphere and oceans. Venus, on the other hand, may have had a primordial excess of hydrocarbons, which would lead to an atmosphere of carbon dioxide and smog, and to oceans of oil.

That four such diverse pictures of the surface conditions on Venus could seriously be entertained at the same time is an indication of the paucity of real data prior to 1956. At that time radio observations at the Naval Research Laboratory disclosed the fact that Venus is very hot, about 300 degrees Celsius.

As is usual in such cases, this new datum raised more questions than it answered. Why was Venus so hot? The explanations offered agree that some mechanism allows solar energy to enter the atmosphere easily but prevents the resulting heat from radiating away as easily. They disagree markedly on what this mechanism, called the greenhouse effect, might be. The mechanism must be similar to the familiar florist's greenhouse which traps solar energy because light radiation penetrates the glass but longer-wave heat radiation cannot escape.

One greenhouse theory suggests that the carbon dioxide of the atmosphere is responsible. However, it would require very high atmospheric pressure before the temperature could rise to 300 degrees C. Another current theory is the so-called *aeolosphere theory*. According to this view, solar energy is captured in the form of the mechanical energy of great dust storms. These storms lash the surface continuously, generating the needed heat. This heat is retained, so the theory goes, by the insulating action of the dust clouds.

The voyage of Mariner II was the first successful attempt to bring scientific instruments to the vicinity of Venus—or any planet—and the journey was a great technological feat. The craft had, in effect, been launched three times: first from the surface of Earth, then from a "parking" orbit around the Earth, and finally from an orbit of the Sun, 9 days and 1.5 million miles from the Earth, where it was put through a maneuver to place it in a new solar orbit. Throughout most of the flight it maintained a rigid orientation with respect to the Sun and also with respect to the Earth. Mariner II spent 109 days, on its way to Venus, gathering scientific data in the void between the two planets. The interplanetary magnetic fields, the solar wind, and dust particle density were monitored almost continuously.

A magnetometer aboard Mariner II showed no measurable magnetic field due to Venus at the distance of closest approach, 36,500 kilometers. This was taken to signify that Venus rotates slowly or not at all. It must be noted, however, that earlier radar measurements demonstrated that the period of Venus' rotation is approximately 250 days, retrograde. The relationship between magnetic field and rotation is complicated by the fact that Mars also has no magnetic field, but rotates as rapidly as does the Earth.

One of the instruments Mariner II carried to Venus was a microwave radiometer, operating at a wavelength of 19 millimeters. This instrument was designed to measure the variation of apparent surface temperature across the disk of Venus. Earlier, Earth-based measurements of the microwave radiation from Venus had led to temperature estimates of 300 degrees Celsius. These measurements, however, engendered lively debate as to whether the radiation originated from the surface or from the atmosphere of Venus. The two possibilities lead to very different surface temperatures.

Apparently, the Mariner II radiometer settled this question by measuring a definite limb darkening of the radiation—that is, the radiation was more intense at the center of the disk than at the edges, an effect to be expected if the radiation comes from the surface and is partially absorbed by the atmosphere. If the radiation source were in the atmosphere, however, the reverse would be true; then the edges of the disk would be brightest. Thus, the Mariner II results show that it is the surface which produces the radiation, and that the temperature there is about 450 degrees C.

Mariner II also carried an infrared radiometer operating at a wavelength of 10 microns. This instrument measured the temperature of the cloud tops, and found it to be about minus 35 degrees C. Enough resolution was provided to be able to detect any break in the cloud cover, but none was observed.

With this information it has been estimated that the midnight temperature on the surface of Venus is probably greater than 300 degrees C. At noon the temperature probably rises to 700 degrees C. Possibly it is cooler at the poles, perhaps as low as 200 degrees C. These extreme temperatures make it unlikely for any kind of life to exist on the surface of Venus. Water is thought to be necessary for the development of life, but there can be no water at such temperatures. Perhaps some liquid water can exist at the poles, if the atmospheric pressure is enough to prevent the water from boiling away. Even so, it is hot enough to destroy the complex organic compounds which are the basis of Earth's life.

Life on Earth, however, is extraordinarily varied. It is so adaptable to the extremes of Earth's environment that one cannot completely exclude its existence on Venus. The surface may be too hot, but perhaps microorganisms can exist in the cooler parts of the atmosphere, kept airborne by turbulent winds, or perhaps on mountain peaks that reach to sufficient heights.

Each new scientific instrument, when applied to Venus, has rewarded its user with greater and often surprising knowledge. So it has been with modern radar. The first successful astronomical radar experiment was performed in 1946, using the Moon as a target. Because of its much greater distance, even at closest approach, Venus is ten million times harder to detect by radar than is the Moon, and the first successful observations of Venus with radar occurred in the spring of 1961. So far, four planets—Mercury, Mars, Jupiter, and Venus—have been observed by radar.

The basic measurement that radar makes is that of the power of radio waves reflected from the target. If an impulse of waves is beamed at Venus, for example, the earliest part of the echo will have been reflected from the front cap of Venus. As time continues, the impulse will travel across the planet and the echo power will have been reflected from successively greater distances from the front. Thus measuring the echo power as a function of time delay reveals the reflectivity of Venus as a function of distance from the closest, or sub-Earth, point.

Time-delay measurements can also be used to measure the distance to Venus. Such measurements were first made in the spring of 1961 and resulted in an extraordinarily accurate value for the astronomical unit. The value was more accurate, by a factor of nearly one thousand, than the best measurements made before that time. Thus, Venus has fulfilled the old prophecy of Halley, but in a different way.

A second basic phenomenon accessible to radar is Doppler shift—that is, if a pure monochromatic tone is beamed at Venus, the echo will no longer be a pure tone. There will be a frequency shift, caused by any relative velocity between Venus and the radar station, and a frequency broadening caused by the rotation of Venus. As Venus spins, echoes from the approaching limb will return with a higher frequency, and those from the receding limb will be lower. Echoes reflected from Venus will, therefore, be spread into a spectrum of frequencies. Thus analysis of the received signal into its frequency spectrum reveals how the reflectivity of the planet varies from the approaching to the receding limb.

It should be pointed out here that the narrowest radar beam width possible today is still very much larger than the angle subtended by Venus, even at closest approach. Because of this fact, the planet is more or less uniformly illuminated by the radar beam. In order to probe specified areas on the surface of Venus, therefore, it is necessary to make use of the time-delay and Doppler-frequency-shift methods just discussed.

Time-delay and frequency-shift methods may be combined to yield a type of two-dimensional map of the surface of Venus. First, the echo which originates from a specified distance (time delay) is isolated from the rest of the signal. Then a spectrogram of this part of the echo shows the variation of reflectivity along this specified circle on Venus. Thus an area on Venus of, say, unusually high reflectivity would be observed as an echo with a certain time delay and a certain frequency shift.

The history of radar studies of Venus is quite recent. It got off to a false start in 1958 and 1959 when the ubiquitous and overwhelming background noise was identified as possible signals. In the spring of 1961, however, unequivocal detections of Venus were made at five radar observatories scattered throughout the world. That series of experiments provided the measurement of 149,598,000 kilometers for the astronomical unit. It established the radar cross section of Venus at about 11 percent of its geometric cross section, and it demonstrated that the surface of Venus is somewhat smoother than the Moon. In addition, it was found that Venus rotates much more slowly than does the Earth. At the inferior conjunction of Venus in 1962, when more radar capability was available, it was established that Venus' rotation is retrograde, about one revolution per 250 Earth days. All of the Sun's planets revolve around the Sun in the same direction; in addition, the major planets all rotate in this same direction—except Venus. This single exception is important because there is no adequate theory to account for it.

The inferior conjunction of 1964 found Venus under the observation of seven radar stations. Time-delay measurements improved the accuracy of the astronomical unit still further and, in addition, improved the other elements of Venus' orbit. Spectral and time-delay analysis produced the first rudimentary map of the Venusian surface. Several features, or areas of high reflectivity, were observed and these features may perhaps be mountains.

When Venus next returned to the neighborhood of Earth in January of 1966, these same features were observed again. Because they returned in their expected position, it was established that they are fixed to the Venusian surface and do not float in seas, for example, or in clouds. Timing these features provided a most accurate rotation rate for Venus: the period of rotation is 243 days with an error of less than a half day. The axis of rotation is almost (but not exactly) perpendicular to the plane of Venus' orbit.

This is a most remarkable rotation period because it is in synchronism with the Earth. Consider, for example, a Venusian observer at the sub-Earth point at the time of an inferior conjunction with Earth. It would be midnight for this observer and he would see the Earth in his zenith. Then, for him, the Sun would rise in the west and set in the east five times. At exactly the fifth midnight (575 Earth days later), the Earth would be back in his zenith. Thus we are led to ponder the twin anomalies of the rotation of Venus: synchronism with Earth and a retrograde direction.

• 18

Mercury and Pluto

HYRON SPINRAD

Mercury and Pluto, at opposite extremes in the solar system, nevertheless have a few common physical properties. They are both small planets, both are remarkable in having highly eccentric orbits, and probably both have mean densities like that of our Earth. They both must be nearly airless worlds, but they do not now strike us as drearily hot and cold, as we had previously imagined.

Mercury was long thought to have the dual peculiarities of an extremely hot and permanent day side, and a perpetually night hemisphere. The day side would be at about 600 degrees Kelvin (320 degrees Celsius) and the night side just above absolute zero (less than 50 degrees K). We now think that the night side of Mercury is much warmer than this and, indeed, quite moderate in temperature. This recent revision of our ideas stems partly from the newly determined, faster rotation of Mercury, which I shall discuss later.

In size, mass, and density Pluto belongs to the group of terrestrial or minor planets as against the gaseous, gigantic major ones (Jupiter, Saturn, Uranus, and Neptune). Pluto certainly is a cold planet, but it is of some special interest because its density may be rather high. If it turns out that its density is low, its discovery will sound more and more like a peculiar accident because the search for Pluto was based on its perturbation of the orbit of Neptune, which would require an appreciable mass. Still another fascinating but hypothetical question regarding Pluto has to do with its possible origin as a satellite of Neptune: such an origin could be unique in the development of the solar system.

The history of the two planets could hardly be more different. Mercury, the messenger god to the ancients, has been known for thousands of years. Pluto was only discovered in 1930 as a result of a systematic search for a new outer planet. Astronomical observations of these planets have increased appreciably during the past three years so that the classical descriptions of Mercury and Pluto are considerably out of date, and many topics I shall discuss here are still controversial.

MERCURY

Mercury, the closest planet to the Sun, has always been hard to observe because of its very proximity to the Sun, whose brilliance generally masks the light reflected by the planet. Ancient observers found it both as an evening object, seen low in the west after sunset, or as a morning "star," observed briefly in the east prior to sunrise. For a long time, failing to recognize these as the same heavenly body, the Greeks called the former Mercury and the latter Apollo. Most present-day observations of Mercury in the radio and optical regions of the spectrum are made in broad daylight. However, visual observations, even under the best conditions, remain quite difficult because Mercury is usually about 80–100 million kilometers from us at its closest and presents a disk of apparent diameter of only about 10 seconds of arc at that distance.

Mercury has an orbit that deviates considerably from a circle around the Sun. Mercury's mean solar distance is 0.39 astronomical units or about 60 million kilometers, but the eccentricity of its ellipse is 0.206, so that at perihelion it is only some 47 million kilometers from the Sun and at aphelion about 70 million kilometers distant. Mercury speeds around the Sun in 88 days, and so its orbital velocity is high, averaging 48 kilometers a second.

The planet is our smallest, with an equatorial radius of about 2,400 kilometers, less than 0.4 the Earth's radius. These data seem firmly established. However, a major revision has recently taken place in our knowledge of the rotational period of Mercury. Rather marginal, old visual observations of very indistinct markings on the surface suggested that the period of rotation was exactly equal to the period of revolution around the Sun, 88 days. In this case Mercury would always keep one face permanently toward the Sun in much the same way that the Moon's rotational period is synchronized with its revolution about the Earth.

The concept of synchronism survived in the literature on Mercury for over sixty years. But recent radar data indicate a substantially faster rotation.

In early 1965 Gordon Pettengill and Rolf Dyce of the Arecibo Ionosphere Observatory made radar observations of Mercury and derived a value for the rotation period of 59 ± 5 days. This value came from an examination of the amount of rotational Doppler broadening suffered by an initially sharp radar pulse sent from the Earth and bounced off the entire surface of Mercury. The faster a planet rotates, the larger will be the difference of the Doppler shift of returning echoes from the two oppositely moving limbs of the rotating planet.

The radar results seem to argue decisively for a shorter rotation period, which raises the following question. Is there any way to reconcile the modern data with the old visual observations? Apparently this can be done, and W. E. McGovern, S. H. Gross, and S. I. Rasool point out that a comparison of visual drawings of Mercury's surface markings does not necessarily indicate synchronous rotation of the planet as the only possible solution. A number of rotation periods are possible and, fortunately, only one is consistent with the 1965 radar results. That rotation period is 58.4 ± 0.5 days.

The 58-day rotational period now suggested may be explained dynamically by taking into account the differential tidal forces exerted by the Sun at the aphelion and perihelion points of the orbit of Mercury. Because Mercury has a very elliptical orbit, these forces will be maximized at perihelion; in fact, a rotation period exactly two-thirds of the orbital period may be the most stable rotation period for Mercury. More empirical and theoretical work is necessary here, however, before a definitive answer can be had.

While the previous theory implied permanent sunlit and dark hemispheres, the non-synchronous rotation of Mercury now assures us that all parts of the planet see the Sun. The length of the Mercurian day is roughly equal to 6 months on the Earth. Temperatures on Mercury's sunlit hemisphere ought certainly to be high, for on the average each square centimeter of the planet's surface receives about seven times the amount of solar energy reaching the Earth. Infrared observations made at Mount Wilson Observatory in the 1930's indicated a temperature of about 600 degrees Kelvin (320 degrees Celsius) for the sunlit hemisphere, but the dark side was too low to measure with these other, relatively insensitive heat detectors. Recent radio observations—passive measurements of the heat radiated by Mercury at microwave wavelengths—indicate that the dark hemisphere may be rather warm, possibly as high as 300 degrees K (+27 degrees C). Nonsynchronous rotation would be necessary to explain this observation, as models based on a permanently dark hemisphere—one which never sees the Sun—cannot yield values close to the currently estimated temperatures. The dark side might be heated by the circulation of a thin atmosphere, and I shall discuss this point later. However, it will still be a few years before ground-based temperature measurements of Mercury in the infrared and radio regions can be obtained with sufficient precision to settle the old but still outstanding arguments. For example, the cooling rate of the Mercurian surface at night can tell us something about its composition and porosity; for this we need radio observation of the dark side at different times and different wavelengths.

Although the atmosphere of Mercury was long regarded as nonexistent, it is now a topic for debate and observation. The historical skepticism regarding Mercury's atmosphere is based upon the small surface gravity and high day-side temperatures prevalent on the planet. Under these conditions most atoms and molecules achieve high velocities and can escape from the gravitational field of Mercury in times shorter than the age of the solar system (some 4–5 billion years). These theoretical expectations are supported by the optical reflectivity properties of Mercury, which are very much like our completely airless Moon.

Early spectroscopic searches for a Mercurian atmosphere failed. Yet infrared spectroscopy of this sort was, of course, successful in detecting carbon dioxide in the atmosphere of Venus and methane (CH_4) and ammonia (NH_3) in the major planets. It was therefore generally believed that, if it exists, the atmosphere of Mercury is composed of rare, inert gases—perhaps argon, which could originate from crustal radioactive decay of potassium. Argon is heavy and is not observable spectroscopically in planetary atmospheres.

Recently, astronomical thinking on this matter has been rudely awakened by the discovery of weak carbon dioxide bands in the infrared spectrum of Mercury by V. I. Moroz in the Soviet Union. The observation is difficult due to the presence of carbon dioxide (CO_2) in the atmosphere of the Earth. But if confirmed, Moroz' work will prove beyond doubt that Mercury is not an airless world. The amount of CO_2 is not immediately obtainable from Moroz' observations; attempts in the United States to find other, weaker carbon dioxide features have failed. In principle, by observing carbon dioxide bands of considerably different intrinsic strengths, we should be able to deduce both the total number of CO_2 molecules above the Mercurian surface and the surface pressure, which may be dependent upon other gases than CO_2. At this time only crude limits on the surface pressure exist: if Moroz's observations are correct, then the atmospheric pressure at the surface of Mercury may exceed 4 millibars, or 0.4 percent that at sea level on the Earth.

Polarimetric observations of Mercury may also indicate a weakly polarizing atmosphere there, but their interpretation is clouded by an uncertainty about the degree of visual light polarization to expect from the Mercurian surface alone. With certain assumptions about the underlying surface polarization, A. Dollfus in France computed a surface pressure of approximately 1 millibar for Mercury. In summary, present empirical evidence points, rather insecurely, to a tenuous atmosphere on Mercury.

How does the planet retain gases like carbon dioxide or argon despite the escape of molecules to space from the hot illuminated hemisphere? The temperature of Mercury's upper atmosphere is the key to molecular escape. The temperature at the "escape level" in our own atmosphere is approximately 1,500 degrees Kelvin; if Mercury has a very hot upper atmosphere (say 5,000 degrees Kelvin), then carbon dioxide and argon would be quickly lost, and one would not expect even a trace of a primitive Mercurian atmosphere to remain. At 1,000 degrees Kelvin carbon dioxide will escape slowly; to observe it now we probably would require a large amount of outgassing of CO_2 from the interior of the planet; and this possibility is difficult to appraise at the present time.

PLUTO

Pluto, the most distant planet known, was discovered by a systematic astronomical search. The search for Pluto, or rather "Planet X," was initiated mainly through the efforts of Percival Lowell. Lowell's studies of the motions of Uranus and Neptune suggested to him that a planet of sizable mass coursed well beyond Neptune's orbit. Fortunately, Lowell realized that to make feasible a survey for a faint outermost planet over large areas of the sky, photography of the ecliptic had to replacé visual examination of stars. The Lowell Observatory, in Flagstaff, Arizona, was dedicated to this work. For several years prior to Lowell's death in 1916 the search went on, but it produced only chance discoveries of many asteroids and variable stars.

In 1929 the search was begun again by Clyde Tombaugh at Flagstaff, this time with a new telescope, a 13-inch refractor. The photographs with this telescope covered an area of the sky 12 by 14 degrees. Consequently, many, many star images were recorded on each plate—sometimes as many as 500,000. Could Planet X be found among all these other images? The techniques adopted took advantage of the expected motion of Pluto in the sky over intervals of a few days. Asteroids moved farther, stars not at all. After only a half-year of plate examination, Tombaugh found the image of Pluto on three photographs, and the discovery was announced on March 13, 1930. The planet was faint and showed no visible disk under close visual examination. Lowell had expected Planet X to be as large as Neptune, and so there was some suspicion at the time that the main outer planet still might be undiscovered.

Physical data on Pluto remain very uncertain even now. The most basic quantities—the mass, radius, and mean density of the planet—are subject to considerable improvement.

The mass of Pluto must be derived from its small gravitational effect on the orbits of Neptune and Uranus, for Pluto has no satellite. Lowell's initial calculations suggested a planet like Uranus or Neptune, larger and more massive than the Earth. More recent calculations seem to throw doubt on this: the uncertainties in very old observations of Neptune suggest that the mass of Pluto could be like the Earth's or even much smaller. If so, then the search for Pluto and its discovery would become a matter of happy circumstance.

The diameter of Pluto is small, and attempts to measure the planetary disk directly at the telescope have been very difficult. The visual observations of G. P. Kuiper lead to a rather approximate diameter of 0.23 second, or 5,900 kilometers—about 40 percent that of the Earth. If the mass of Pluto were about that of the Earth, then the mean density would be about 50 grams per cubic centimeter, an impossibly high value. It is therefore important to know whether the diameter could have been underestimated by Kuiper.

A recent novel contribution to this subject has been the study of Pluto's diameter by star occultation. This work has been undertaken cooperatively by several astronomical observatories, under the direction of I. Halliday in Canada. The idea is simple enough: a distant planet like Pluto, although presenting only a tiny disk some 0.2 second in diameter, will occasionally eclipse (or occult) background stars as it moves slowly through the sky. Prior to occultation we may accurately observe the brightness of the star plus Pluto; during the occultation, we see only the brightness of Pluto, which shines dimly by reflected sunlight. To measure the diameter of Pluto we need to measure the duration of the occultation; the speed in the orbit is well known. If such observations are made successfully at several observatories at different parts of the Earth, we can be sure of the exact trajectory of Pluto as it occults a star, and the diameter of Pluto can be determined from the geometry of the occultation.

The difficulty with this method lies in predicting the orbital path of the planet with enough precision to tell observers in advance whether Pluto will pass closer than 0.5 second or so of an observable (usually very faint) star. In fact, the positions of likely stars need redetermination.

Halliday's current program has recently yielded some important results. First of all, we may now estimate the expected frequency of occultations. For the 1963–64 observing season, Pluto passed sixteen stars within 10 seconds of arc; in 1964–65, the number was nine stars. Statistically, then, Pluto should pass within about 1 second of arc of a star each year, the stars being brighter than magnitude 17 (or some 10,000 times fainter than naked-eye vision on a clear, dark night).

A possible occultation of a faint (magnitude 15) star was predicted for April 28, 1965, and numerous observatories agreed to attempt photographic and photoelectric observations of Pluto and the star that evening. Unfortunately, no occultation was observed. The southernmost stations in the United States probably had a very close miss. But we can say with considerable assurance now that Pluto is smaller than 6,800 kilometers in diameter, increasing our confidence in Kuiper's visual measurement.

One other physical fact known about Pluto is its rotation rate. The day on Pluto is a little over 6.3 Earth days. This is rather slow rotation for a planet; only Mercury and Venus are slower, and here tidal effects of the Sun might be responsible. Kuiper has suggested that the slow rotation of Pluto could be accounted for by assuming it was initially a satellite of Neptune (like Triton) which later somehow escaped. The 6-day rotation period is compatible with a satellite orbital period about Neptune; in fact, it is only slightly larger than Triton's present orbital period about Neptune. This interesting speculation is very difficult to check at the present.

No atmosphere has been found for Pluto spectroscopically. It is a terribly faint object for suitable observation, but the expected low temperature is likely to freeze out gases like H_2O, NH_3, CO_2, and perhaps even CH_4. Not much of a gaseous atmosphere would remain above the snows and ices expected at a 50 degrees Kelvin (−220 degrees Celsius) temperature. The amount of solar radiation reaching Pluto is less than a thousandth of that falling on each square centimeter of the Earth's surface. Nevertheless, surprises have happened in planetary astronomy, so the case should not be considered permanently closed.

Many of the perplexing questions concerning Mercury and Pluto are not going to be easily answered. The techniques of ground-based optical and radio astronomy are somewhat limited, and although new information will be forthcoming, certain observations—perhaps of the presence of an argon–carbon dioxide atmosphere for Mercury or even an accurate mass for Pluto—will probably come from spacecraft.

No probes are yet scheduled to fly past Mercury. However, the duration of the trip would be moderate (less than a year) and temperature control would be possible. Certainly solar cells could be small, for there is plenty of solar power available. A Mercury fly-by or orbiter could perform direct photography of the surface and an occultation experiment like that on Mariner IV could measure a possible ionosphere of charged particles near Mercury and the density of its thin atmosphere. A fly-by could also measure the expected small magnetic field of Mercury. There seems to be no way at all to do this from the Earth's surface. The magnetic field measurement may shed some light on the history of Mercury's rotation period.

Space voyages to Pluto are for the distant future. The travel time would be decades, and the use of solar power out of the question. Probably a nuclear power plant would be required on the probe. A view of the icy surface of distant Pluto from a close vantage would be exciting enough, but to gather the results in this case will require great patience. If and when such a venture is attempted—involving a voyage of more than 6 billion kilometers—it would be well to plan it so that information could be obtained from other heavenly bodies on the way.

19

Jupiter and Saturn

RAYMOND HIDE

The main goal of planetary science is the eventual understanding of the evolution of the solar system and the circumstances of its birth. Many kinds of investigation are involved in acquiring the knowledge necessary to achieve this goal, including purely theoretical studies as well as ground-based observations with telescopes and other instruments. Recent advances in space technology have given scientists a foretaste of the exciting and rapid progress to be expected in the future from the use of instruments mounted on space vehicles and on the surface of the Moon.

Nine tenths of the material of the solar system outside the Sun itself goes into making up two of the planets, Jupiter and Saturn. Most of the angular momentum of the solar system is tied to Jupiter's motion of revolution about the Sun, and Jupiter is the only planet, apart from the Earth, known to possess a magnetic field of its own and to be surrounded by Van Allen-type belts of electrically charged particles. Jupiter and Saturn are associated with two of the curiosities of the solar system, Jupiter's Great Red Spot and Saturn's rings; and both, as compared with the other planets, possess rich satellite systems. But perhaps the main reason for regarding the study of Jupiter and Saturn as central to planetary science is that, unlike the other planets (including Earth), Jupiter and Saturn may have the same chemical composition as did the primordial material of the solar system. Knowledge of this composition is of cardinal importance for planetary science and cosmology.

JUPITER

Jupiter is the largest of the nine planets that revolve around the Sun. Its diameter is 140,000 kilometers, eleven times that of the Earth, and its mass is 2.5 times that of all the other planets put together. It is not inappropriate, therefore, that this great planet should be named after the king of the gods of Mount Olympus.

Jupiter is the fifth planet in order of distance from the Sun, which it circles once every 11.8 years in an orbit lying between the orbits of Mars and Saturn. The planet can profitably be observed by the Earth-bound astronomer for about 10 of the 13 months that elapse between successive conjunctions with the Sun. At conjunction, when Jupiter and the Earth are at opposite sides of the Sun, the two planets are separated by a distance of 965 million kilometers, or 6.5 times the average distance of the Earth from the Sun (a distance often called the astronomical unit). At opposition, when both planets are on the same side of the Sun, they are then separated by 4.0 astronomical units. At its faintest, when its apparent diameter is 32 seconds of arc, or 1/57 that of the Moon, Jupiter is a little fainter than Sirius, the brightest star in the sky. At its brightest, when its apparent diameter is 50 seconds of arc or 1/36 that of the Moon, Jupiter is twice as bright as Sirius. With the exception of Venus, and occasionally Mars, Jupiter is the brightest planet of all.

Jupiter possesses twelve satellites, more than any other planet. A good pair of binoculars will reveal the four largest ones, first discovered by Galileo in 1610. These are Io, Europa, Ganymede, and Callisto, and they revolve around Jupiter at distances ranging from 6 to 27 times the radius of the planet. Only in much more powerful optical instruments can the eight remaining satellites of Jupiter be seen. One of these satellites, Jupiter 5, lies closer to the planet than the Galilean satellites, at a distance equal to 1.3 times the radius of the planet. The seven remaining satellites revolve about the planet at distances ranging from 165 times the radius of the planet (Jupiter 6) to 340 times (Jupiter 9) and thus lie beyond the Galilean satellites.

The diameter of the smallest Galilean satellite, Europa, is 0.84 that of the Moon, or 0.23 that of the Earth. The diameter of the largest Galilean satellite, Ganymede, is 1.46 that of the Moon, and very slightly bigger than that of the planet Mercury. By comparison, the eight remaining non-Galilean satellites are very tiny indeed, ranging from Jupiter 12 with a diameter of about 12 kilometers or only 0.0035 that of the Moon, to Jupiter 5, about 150 kilometers in diameter. It has been suggested that some of these tiny satellites may have been captured by Jupiter from the belt of minor planets or asteroids that revolve around the Sun in independent orbits lying between those of Mars and Jupiter.

Jupiter reflects back into space as much as 0.44 of the sunlight incident upon it. In this property Jupiter is six times as effective as the Moon and almost as effective as the planet Venus. The reflection takes place at the top of opaque clouds suspended in Jupiter's atmosphere of hydrogen, helium, and methane gases. These clouds, which are probably composed mainly of ammonia crystals, shroud the underlying planet from view.

The broad features of Jupiter's visible disk are revealed by telescopes with apertures as small as 10 centimeters. That the disk of the planet is not quite circular is readily perceived in such a telescope. This oblateness is due to the rapid rotation of the planet—once in less than 10 hours (the shortest rotation period of any planet)—which gives rise to centripetal forces that increase the diameter of the planet at the equator and flatten the poles. The observed oblateness is significantly less than it would be for a planet of the same density throughout, showing that Jupiter is highly condensed towards the center—much more, in fact, than the Earth.

Larger telescopes with apertures exceeding about 30 centimeters reveal many of the details of markings on the visible disk that are only dimly seen in smaller instruments. The most prominent markings are the bright cloud zones, of which there are usually about seven or eight. These zones run parallel to the equator and are separated by darker belts. The belts and zones are not entirely regular: dark patches often appear on the bright regions and bright patches on the dark regions, and the boundaries between the belts and zones often take on a serrated shape.

The most striking marking of all is the Great Red Spot, which occurs in Jupiter's southern hemisphere. This object was first observed by Robert Hooke and Giovanni Cassini during the second half of the seventeenth century. It is elliptical in shape, having its long axis along latitude 22 degrees. It occupies 24 degrees of longitude and 12 degrees of latitude, and thus covers an enormous area, roughly equal to that of the Earth's surface, though only about 1 percent of the surface area of Jupiter.

Information on the rotation of Jupiter has been obtained by studying the motion of markings on the visible disk. Though not always the same, the rotation periods thus measured are always several minutes under 10 hours. They vary with latitude in a complicated way which is unsymmetrical about the equator, and there is no doubt that these variations in rotation period are manifestations of atmospheric motions. There is an eastward current of about 100 meters per second extending from the equator to latitude 7 degrees, approximately, in both hemispheres. Atmospheric wind speeds in higher latitudes are about 2 meters per second.

The rotation period of the Great Red Spot has undergone variations of as much as 30 seconds. These variations have been taken by some theoreticians as evidence that the Spot is an object floating in Jupiter's atmosphere. Fortunately, there is no need to invoke the existence of such an unusual object, which would be unlikely to remain in the same latitude. Hydrodynamic theory shows that, owing to the rapid rotation of the planet, quite a shallow "topographical feature" of the surface underlying Jupiter's atmosphere could disturb the atmospheric winds at all levels above it, giving rise to a columnar disturbed region. The Great Red Spot is probably the other end of this column.

It is the tilt of the plane of the equator of a planet to the plane of its orbit about the Sun that gives rise to seasonal variations with the orbital period—a year in the case of the Earth. This tilt for Jupiter is only 3 degrees, which is much smaller than for any other planet, and seasonal effects should, therefore, be slight. Because Jupiter is the only planet whose orbital period is close to that of the well-established, roughly 11-year cycle of surface activity on the Sun, an obvious complication arises in distinguishing true seasonal effects on Jupiter from effects that may result from interactions between the Sun and Jupiter.

Although we do not yet possess very accurate information about the color of Jupiter, there is no doubt that white, gray, yellow, orange, red, brown, and occasionally blue and green regions occur. Though the Great Red Spot is on the average redder than the rest of the planet, it has occasionally appeared to be gray. It has been speculated that Jupiter's colors are due to traces of certain metals such as sodium or potassium in the ammonia crystals that make up Jupiter's reflecting clouds, or to the presence of certain free radicals produced by electrical storms in Jupiter's atmosphere or by ultraviolet radiation from the Sun.

According to measurements of infrared radiation emitted by Jupiter, the surface of the planet is quite cold, about minus 120 degrees Celsius. Nevertheless, the surface is significantly hotter than it would be if solar radiation were the only form of energy reaching it. It has been estimated that Jupiter's visible surface receives at least as much energy from the interior of the planet as it receives from the Sun. This calls for an internal energy source about a hundred times more powerful per unit mass than that of the Earth. This source could be accounted for quite readily in terms of the conversion of gravitational energy into heat if Jupiter is still slowly collapsing or accreting matter from outside. Alternatively, it would be necessary to postulate an extraordinary degree of radioactivity in the small fraction—probably about 2 percent—of Jupiter's mass that is made up of heavy chemical elements.

Variations in temperature across the visible disk are small, only a few degrees Celsius, and are consequently difficult to measure accurately. It will be important to establish whether or not color variations are associated with these temperature variations.

The mass of Jupiter and some information on its distribution with depth within the planet can be derived, using gravitational theory, from the motions of its satellites. The average mass of Jupiter per unit volume, or mean density, is 1.334 grams per cubic centimeter. This value, though comparable to those of the other major planets (Saturn, Uranus, and Neptune) and remarkably close to that of the Sun, is only about one fourth the densities of the comparatively small terrestrial planets (Mercury, Venus, Earth, and Mars). This low mean density is evidence that the main constituents of Jupiter are the light elements, hydrogen and helium. Owing to Jupiter's high surface gravity (2.6 times that of the Earth) and its low surface temperature, thermal motions of even the lightest of the constituent molecules of the gases of Jupiter's atmosphere would never have been so great as to overcome the pull of gravity. The relatively high mean densities of the terrestrial planets are usually taken as evidence that, unlike Jupiter, these planets are too small to be able to retain gases of the light elements in their atmospheres.

Owing to its great mass and comparatively low surface temperature, Jupiter is possibly a specimen of the primordial material of the solar system. It is of cosmogonic interest, therefore, to obtain knowledge of the chemical composition of the whole planet. In principle this can be done by building theoretical models of the planet. Thus, model planets are constructed by using the laws of physics to calculate how the density, pressure, and temperature would vary with distance from the center for a mass of material equal to the known mass of Jupiter. Each model is characterized by certain assumptions, including those made about the chemical composition. There are many sources of error in the calculations, including uncertainties about the behavior of matter at the very high pressures (many millions of atmospheres) that prevail throughout most of the planet. Models that are incompatible with what is known about the density distribution within the planet from studies of the flattening of Jupiter and of variations of the orbital motions of the innermost satellites are rejected.

The most plausible of the remaining models suggests that, by mass, 80 percent of Jupiter is hydrogen, 18 percent helium, and 2 percent heavier elements, that it possesses a deep, well-stirred atmosphere, and that, owing to the high prevailing pressures, it has metallic properties from the center out to about 0.8 of the radius of the planet. Although these models give the pressure and density distribution within the planet fairly well, they are still incapable of predicting accurate temperatures. For this reason, we do not yet know whether Jupiter's deep interior is solid or fluid.

The invention several decades ago of the radio telescope has added a new dimension to astronomy. The radio astronomer studies electromagnetic radiation from astronomical bodies on very much longer wavelengths than those characteristic of visible light and infrared radiation. Electromagnetic radiation from Jupiter on wavelengths ranging from decimeters to decameters has many fascinating and quite unexpected properties. On these wavelengths Jupiter is one of the brightest radio sources in the sky, emitting much more energy, by several powers of ten, than the thermal radiation to be expected from a body of such size and distance with a surface temperature of minus 120 degrees Celsius. The decameter radiation is emitted in intermittent bursts from relatively localized sources. The frequency of occurrence of the bursts has been shown to depend on the position in its orbit of the satellite Io.

Jupiter's non-thermal radiation was discovered, by accident, as recently as 1955. Since that time, intensive studies of the detailed properties of this radiation have provided some very exciting results. Although no entirely satisfactory theory of the radiation has yet been put forward and the Io effect is particularly mysterious, there seems to be no doubt that most of the radiation originates in belts of electrically charged particles surrounding Jupiter. The electrons in these belts are a thousand times more energetic than those trapped in the Van Allen belts surrounding the Earth. The magnetic field required to keep these charged particles in the vicinity of Jupiter must come from within the planet. The strength of this magnetic field at the visible surface of Jupiter may be about a hundred times that of the surface magnetic field of the Earth.

Jupiter's magnetic field is probably due to a self-maintaining magnetohydrodynamic dynamo mechanism operating somewhere within the planet. According to the theory of such dynamos, motions of an electrically conducting fluid can, under certain circumstances, interact with a magnetic field in such a way as to produce the electric currents required to maintain the magnetic field against agencies tending to destroy it. The possibility that the magnetic field of Jupiter is produced by dynamo action in the lower reaches of its atmosphere is consistent with the evidence available, although it must be admitted that this evidence is scanty and that new observations might change the picture considerably. Contributions to Jupiter's magnetic field might also arise from deeper down if the planet has a fluid core.

The theoretical discussion of Jupiter's internal motions has only just begun. This discussion promises to progress rapidly in the near future as radio astronomers learn more about the properties of Jupiter's radio sources and fluid dynamicists extend the theory of the hydrodynamics and magnetohydrodynamics of rotating fluids. Only tentative explanations have been given of Jupiter's banded appearance, rapid equatorial current, Great Red Spot, and general magnetic field. Making sense of these observations without doing violence to the laws of physics is a fascinating study which should add considerably in due course to our knowledge of the interior of the planet. It has recently been shown, for example, that the variable rotation period of the Great Red Spot can be taken as evidence that the magnetic field in Jupiter's lower atmosphere may be twenty times as strong as the field at the visible surface.

SATURN

Saturn revolves around the Sun once every 29.5 years at a distance of 9.5 astronomical units, nearly twice that of Jupiter. It was the most distant planet known in remote antiquity: Uranus, Neptune, and Pluto have been discovered since the Middle Ages with the aid of telescopes and, in the case of Neptune and Pluto, with the further aid of the theory of universal gravitation.

Saturn's diameter of 113,000 kilometers is 0.82 that of Jupiter, and its mass is 0.3 that of Jupiter. Though quite bright, Saturn is fainter than Jupiter by a factor of about ten. Lying in the plane of Saturn's equator is a system of rings surrounding the planet, ranging in diameter from 135,000 kilometers to 270,000 kilometers. These rings are a very beautiful sight even in quite a small telescope. Variations in the brightness of Saturn as seen from the Earth are caused more by the varying angle of presentation of the ring system than by variations in the distance between the Earth and Saturn associated with the movement of the two planets in their respective orbits.

Saturn has nine known satellites, although late in 1966 tentative evidence was announced of a tenth and very tiny satellite in an orbit lying closer to the planet than those of the others. The latter nine revolve in orbits ranging in diameter from 4.1 to 220 times that of the planet and thus lie outside the system of rings. Though in many ways comparable with the satellites of Jupiter, Saturn's moons are somewhat larger. All of them have been given names: in order of distance from the planet, they are Mimas, Enceladus, Tethys, Dione, Rhea, Titan, Hyperion, Iapetus, and Phoebe. Titan has the greatest mass and is second largest in size of all the moons in the solar system. It is, moreover, the only satellite known to have an atmosphere of its own.

In spite of the spectacular appearance of Saturn's rings, their mass is only 1/27,000 of that of the planet and 1/5 that of the largest satellite, Titan. Although they are of great lateral extent in the plane of Saturn's equator, the rings are incredibly thin—only 16 kilometers. They are made up of an enormous number of discrete rocky fragments of irregular shape, each less than a few centimeters in size. Every fragment revolves around the parent planet as a tiny independent satellite.

Certain regions of the ring system have been swept clean of fragments by the gravitational pull of the nearest of Saturn's larger satellites, especially Mimas. Jupiter produces a similar effect on the minor planets, or asteroids, that revolve around the Sun, giving rise to the so-called "Kirkwood gaps" in the asteroid belt.

Although there is as yet no generally accepted theory of the origin of Saturn's rings, it is very likely that they are remnants of a satellite which approached too close to be able to withstand the disruptive action of the planet's gravitational pull.

Saturn is comparable to Jupiter as a reflector of sunlight. Like Jupiter, Saturn is enveloped in dense clouds of ammonia crystals suspended in an atmosphere of hydrogen, methane, and helium, and arranged in belts parallel to its equator. These belts appear to be more regular than those on Jupiter. Spots and other eruptions are relatively infrequent, and nothing quite comparable with the Red Spot on Jupiter has ever been seen on Saturn. Color variations on Saturn are much less pronounced than on Jupiter.

Transits of long-lived spots on Saturn yield rotation periods of 10 hours and 13 minutes near the equator and 10 hours and 40 minutes in mid-latitudes. Thus Saturn, like Jupiter, exhibits evidence of an equatorial current. The corresponding wind velocity is 400 meters per second in an eastward direction, four times faster than that on Jupiter. This result has been interpreted tentatively as evidence that Saturn's atmosphere may be much deeper than Jupiter's.

Saturn's surface temperature, according to infrared measurements, is about the same as Jupiter's, approximately minus 120 degrees Celsius. As with Jupiter, it is necessary to invoke a substantial source of internal heating to account for this surface temperature.

Owing to its rapid rotation, Saturn is oblate, having an equatorial diameter over 10 percent greater than the polar diameter. This degree of oblateness, half again as much as Jupiter's, is greater than that of any other planet.

Saturn's mean density is only 0.715 gram per cubic centimeter, 0.53 that of Jupiter. There is no other planet of such low density. Theoretical studies suggest that Saturn strongly resembles Jupiter in chemical composition and in internal structure.

The radio astronomer who has observed Jupiter finds Saturn a disappointing object. There have been recent reports of fleeting radio bursts, but these are less frequent and much weaker than the bursts from Jupiter and may, in fact, be spurious. Several possible interpretations of these observations have been suggested. Saturn may not possess a magnetic field capable of forming radiation belts by holding electrically charged particles trapped in the vicinity of the planet. Alternatively, if the particles in the radiation belts of the Earth and Jupiter come from the Sun in the so-called solar wind, the lack of radio emission from Saturn could be accounted for by supposing that the solar wind does not reach as far as Saturn. One interesting speculation is that the rings of Saturn may prevent radiation belts from forming.

The study of Jupiter and Saturn is an important area of scientific inquiry engaging the efforts of investigators of diverse backgrounds, schooled in various disciplines. Present knowledge of these two great planets, as outlined in this chapter, raises many interesting questions, which in turn suggest crucial observations and theoretical investigations. Many of the required observations will, no doubt, be obtained with ground-based instruments, especially if large new telescopes become available for constant use in planetary work. But a number of important questions may not be answered with certainty without the aid of space probes capable of making measurements close to Jupiter and Saturn.

Such probes will take several years to reach their destinations and communicate their observations back to Earth, an awesome prospect. In addition to settling existing questions, thus deepening our knowledge of the solar system, these space-probe observations are bound to raise new questions, adding fresh excitement to planetary science.

20

Uranus and Neptune

GERARD P. KUIPER

During antiquity and until well into the eighteenth century, it was assumed that there were seven planets (Greek for "wanderers"): the Sun, the Moon, Mercury, Venus, Mars, Jupiter, and Saturn, after whom the days of the week were named. It was difficult to believe that this would ever change.

The discovery of Uranus by William Herschel on March 13, 1781, "had the surprising effect of utter novelty. The event broke with immemorial traditions, and seemed to show astronomy as still young and full of unlooked-for possibilities."[1] This event occurred at the close of a century noted for its development of calculus and celestial mechanics, which had succeeded in accounting for all observed complexities of planetary motion. It was a development that, moreover, through its demonstrated perfection had suggested a degree of finality.

Uranus was found to possess a nearly circular orbit at an average distance of 19.2 astronomical units from the Sun. This distance was comforting. It confirmed and extended the roughly exponential law of Titius-Bode for planetary distances from the Sun.

---

[1]Agnes M. Clerke, *A Popular History of Astronomy during the Nineteenth Century* (London: Adam & Charles Black, 1902), p. 5.

The discovery of the next planet, Neptune, in 1846, was due not to an accidental observation by an assiduous observer but, instead, to brilliant theoretical prediction, based on perturbations in the orbit of Uranus. The observed motion of the planet Uranus did not seem to follow gravitational theory. New data increasingly showed intolerable departures from Newtonian motion. The concept of the irregularities being due to an unknown exterior body appears to date back to about 1830, but no precise predictions of its location were available until 1845 and 1846. These predictions led to the discovery of the disturbing body on September 23, 1846, by J. G. Galle of Berlin, who used the precise forecasts of position and brightness by U. J. Leverrier of Paris. An independent parallel effort in England led to somewhat earlier observations which, however, remained unreduced until after Galle's discovery had been announced.

The discovery of Neptune led to repeated later attempts to extend the planetary system one more step. The discovery of Pluto at the Lowell Observatory in 1930 had been stimulated by such efforts, although the discovery itself by C. W. Tombaugh was an empirical result based on a well-conceived, very thorough photographic survey of most of the sky. Soon it was found that the mass of Pluto is too small to have caused appreciable effects on the observed positions of Uranus and Neptune. Pluto, in fact, appears to be a body that is not only much smaller in mass but also different in composition from Uranus and Neptune. It resembles, instead, a large satellite, and it seems probable that Pluto was a satellite that escaped from Neptune during the early history of the solar system. This explanation is indicated by the strange nature of Pluto's orbit around the sun, for it intersects the orbit of its neighbor Neptune. Thus, Uranus and Neptune, by their large masses and nearly circular orbits, are the outermost objects now known having unquestioned planetary status.

The effects of the discovery of Neptune on the reputation of science were profound. "By it the last lingering doubts as to the absolute exactness of the Newtonian Law were dissipated. Recondite analytical methods received a confirmation brilliant and intelligible even to the minds of the vulgar, and emerged from the patient solitude of the study to enjoy an hour of clamorous triumph. Forever invisible to the unaided eye of man, a sister-globe to our earth was shown to circulate, in perpetual frozen exile, at thirty times its distance from the sun."[2]

The scientific significance of the planets Uranus and Neptune is heightened by the fact that both are attended by satellites. Uranus has five known satellites, two of which (Titania and Oberon) were discovered by William Herschel in 1787, two (Ariel and Umbriel) by W. Lassell in 1851, and the fifth satellite (Miranda) by the writer in 1947. All five satellites have orbits that are very nearly circular and are situated in a common plane, which at the same time is the plane of the planet's equator. This common plane is not, as one might have expected, nearly coincident with the planet's orbit around the sun, but instead is nearly vertically inclined to it. Thus, as the planet moves about the Sun in its orbital period of 84 years, the satellite plane is seen edge-on from the Sun and from the Earth on two occasions, 42 years apart. Such an edge-on appearance occurred in 1966, with the next occurrence due about 2008. Such appearances obviously offer favorable occasions to measure the precise inclinations of the satellite orbits as well as the oblateness of the planetary disk due to its rotation. Because no spots can be seen on the planet, the period of rotation must be ascertained spectroscopically, by the Doppler shift of the reflected light observed along the planetary equator. In this manner the observers at the Lowell Observatory found a period of rotation of 10.7 hours, or 0.45 days.

Interestingly, the period of rotation of Uranus is similar to that of the giant planets Jupiter (9 hours and 50 minutes, or 0.41 days) and Saturn (10.2 hours, or 0.43 days), both measured at the equator. Jupiter and Saturn have a so-called equatorial acceleration, as is true of the Sun, meaning that the equatorial zones revolve somewhat more rapidly than do the higher latitudes. It is not known how this differential effect is maintained in the giant planets. Nor is it known whether Uranus shares in this peculiar property, because its disk is too small and too dim to allow a ready spectroscopic determination of this effect.

Let us pause to consider this period of rotation of about 10 hours, common to the planets Uranus, Saturn, and Jupiter. This period is rather small for bodies having such comparatively low mean densities. Jupiter's mean density is 1.33 times water, Saturn's

---

[2] *Ibid.*, p. 82.

only 0.71 times, and Uranus' 1.56 times. Consider a set of planets whose mean densities are all equal to that of water but whose periods of rotation vary in length. One finds that there is a limit below which the period cannot go. If the limit were surpassed, the centrifugal force at the equator would exceed the attractive force of gravity, and the equatorial material of the planet would fly off into space. For a planet of unit density this limiting period is roughly 0.14 day or 3.3 hours; for a planet of higher density the limiting period is lower, decreasing inversely as the square root of the mean density. Thus the rotational periods of the three planets are only about three times as long as the absolute minimum compatible with stability. Because in the early period of planet formation there could have been several mechanisms by which the forming planet could have shed an excessive amount of rotational momentum and thereby lengthened its period of revolution, it is not unreasonable to assume that the three planets initially rotated considerably faster than at present and that they, in fact, approached the danger limit. These planets may thus have been surrounded by a disklike or sheetlike gas cloud in which condensation took place. It is precisely this mechanism that is assumed to have given rise to the very symmetrical and flat satellite systems of these three planets, and in the case of Saturn to its ring as well.

Because the inner satellites of Saturn are better known than those of Uranus, owing to their greater brightness and greater proximity to the Earth, I shall consider the Saturn system for purposes of general orientation. The masses of its inner satellites are quite well known, derived from gravitational perturbations of long periods that are evaluated from precise positional measurements. The diameters of the Saturn satellites can also be determined: I found them just measurable with the 200-inch telescope. The mean densities can thus be computed and, for the inner two Saturn satellites, are found to be somewhat less than that of water. The colors of the satellites are white and the reflectivities exceedingly high, close to unity. Apparently these bodies are composed largely of snows, derived from water and ammonia. The two bright rings of Saturn, called A and B, appear to be composed of snow also. This was found by the writer from the infrared reflection spectrum during several series of observations, beginning in 1948. Snow, while white to the eye, has substantial absorptions in the near-infrared, beginning at about 1.5 microns. The thickness of the ring system is probably less than 1 meter and actually probably around 10 centimeters. The inner satellites of Saturn and the ring therefore resemble meteorological condensation products which formed in the original extended atmosphere of the planet. It is very probable that the same is true of the satellites of Uranus. These, too, are brilliant white and their masses, known only approximately, are compatible with the hypothesis that they are composed largely of snows.

The difference between the Saturn rings A and B and the Saturn satellites is merely that the rings are formed inside the so-called Roche limit and the satellites outside this limit. The Roche limit is the distance from the planetary center outside of which solid or liquid masses can combine under the influence of their mutual gravitation; inside the limit this is not possible, owing to the preponderance of the planetary tidal force, which keeps the masses apart. At the limit itself the separating power of the planetary tidal force equals the mutual attraction of two small masses in contact. The location of the Roche limit depends on the bulk density of the condensate. If it equals that of the planet, the Roche limit is 2.4 planetary radii from the planet's center. If the condensate density is less, the limit is farther out. Thus our Moon would break up from the tidal forces of the Earth if it came within 2.9 instead of 2.4 Earth radii from the center of the Earth. The Uranus satellites are outside the Roche limit of Uranus, which allowed them to form. The outer edge of ring A of Saturn is 2.3 planetary radii away, and the first satellite, Mimas, is 3.1 radii away. This indicates that the bulk density of the ring material is quite comparable in density to the planet, or 0.7 times water. It could be slightly less but not appreciably more. This is consistent with the previous conclusions about the consistency of the Saturn rings.

We have now reached a point of extraordinary interest for our general ideas on the formation of the planets and their satellite systems. Why is it that the planet Saturn has rings but that none are found around Jupiter and Uranus? The presence of the Saturn ring and of Saturn's satellite system, beginning almost immediately outside the ring (3.1 planet radii for Mimas), shows that the flat disk of cosmic material extending beyond the Saturn equator was almost continuous from the planet itself to 10 or

20 planet radii outward. Were the original protoplanet disks of Jupiter and of Uranus constituted differently?

In the case of Jupiter the absence of a snow ring can be understood. Jupiter is only half Saturn's distance from the Sun and therefore receives four times more solar energy per unit area. Snows have the interesting property of being very cold compared to silicates. They reflect three fourths or more of the solar radiation received but they are excellent emitters in the infrared, so that they cool themselves by radiation to a temperature of only about 80 percent of that of silicates. Thus, the rings of Saturn should have a temperature of approximately 60 degrees above the absolute zero. At this very low temperature the evaporation rate to a vacuum becomes excessively slow, almost imperceptible even during the long history of the solar system. The rings and the snow satellites of Saturn are thus found to be essentially stable against evaporation losses. This would not be true at the distance of Jupiter. One can show that a thin ring like Saturn's would almost certainly have evaporated during geological time.

A further difference between Jupiter and Saturn appears to exist. The innermost of the four Galilean satellites of Jupiter has a density as high as 4.0 times water, while the other three satellites have densities of 3.8, 2.4, and 2.1, respectively, indicating that Jupiter was a powerful source of heat during the period of satellite formation. It is therefore quite probable that Jupiter at no time possessed a snow ring like Saturn's. These arguments, however, cannot apply to Uranus. This planet is twice the distance of Saturn from the Sun, which would permit the preservation of a ring had it ever existed, while the smaller mass of Uranus compared to Saturn would have favored an even colder envelope. All we can say is that no such ring for Uranus has yet been found and, if present, must be exceedingly dim and rarefied compared to the ring of Saturn.

Although the planet Neptune itself has close similarities to the planet Uranus, as we shall see presently, the two satellite systems stand in marked contrast. Uranus with its five satellites, moving in one plane in a common direction, in nearly perfect circular orbits, and in a plane that is also the equatorial plane of the planet, presents a model of regularity and symmetry. Neptune, with its two satellites, shows a system that is very disorderly.

The inner satellite is Triton, discovered by Lassell in 1846 only seventeen days after the discovery of the planet. It moves in a nearly circular orbit which, however, is tilted 20 degrees to the planet's equator; this causes the orbital plane to precess like a top. The rotation of the planet itself is retrograde, with the period of about 16 hours. The motion of Triton is likewise retrograde, that is, in the sense opposite to planetary motion in general. The second satellite, Nereid, discovered by the writer in 1948, has a direct motion inclined only 5 degrees on the planet's orbit, but the satellite orbit is highly eccentric, its distance to Neptune varying by more than seven times. The period of Triton is 5.9 days, the period of Nereid 360 days. This chaotic mechanical arrangement suggests that something very drastic happened to the Neptune system in its early history. The peculiar orbit of Nereid suggests that, while it was probably formed as a satellite of Neptune, it was temporarily lost by the planet to space and was subsequently recaptured in a direct orbit, related to the Neptune orbit but not to the direction of planetary rotation. This explanation is in common to that for eleven other irregular satellites in the solar system. All of these are assumed to have been first formed in proto-planet envelopes, then lost to nearby interplanetary space as the proto-planet masses decreased and the satellites spiraled outward, and, finally, recaptured in subsequent close encounters. Such encounters had to occur sooner or later, owing to the intersection of the orbits of the planets and their lost satellites around the Sun.

It is probable that Pluto, now an independent planet, was similarly shed by Neptune but did not get recaptured during the life of proto-Neptune. Nereid did get recaptured, but the proto-planet was apparently no longer sufficiently dense to cause its highly elliptical capture orbit to be rounded off to a nearly circular shape. Pluto and Triton are not very different in dimension or, probably, in mass. Triton may be a recaptured satellite, as I have stated, but its very nearly circular orbit suggests an alternative explanation as a second possibility: that Triton was never lost but that its orbit was disturbed by a close passage of an intruder into the satellite system. This intruder may have been Pluto. It could have passed through without itself having been captured, by transferring momentum to Triton at right angles to its orbit. This could have left the orbit essentially intact but for the tilt of 20 degrees.

One final remark is in order on the "obliquities" of the outer planets—the angles between the planetary equators and the orbital planes. This angle is 23.5 degrees for the Earth and 25.2 degrees for Mars. For Jupiter it is only 3 degrees, for Saturn 27, for Uranus 98, and for Neptune 140 degrees. These occasionally high obliquities pose a most interesting dynamic problem that appears to bear on the process of planet formation because under present dynamic conditions these obliquities cannot change appreciably.

The early history of the solar system appears to have consisted of three periods.

1. A period of contraction of a prestellar cloud which led to the formation of the proto-Sun in the center of the system and a flat disk of cold gas and condensate surrounding it. This disk-shaped cloud later broke up a gravitational instability into a system of proto-planets. Each proto-planet gave rise to the formation of a single planet, normally accompanied by a satellite system.

2. The proto-Sun at the center gradually heated up and eventually went through a phase of very high luminosity, even in excess of the present brightness of the Sun. This high-luminosity phase stripped the inner planets of their extensive gaseous envelopes, composed mostly of hydrogen and helium, and appreciably depleted the hydrogen and helium contents even of the planets of intermediate size, Uranus and Neptune. The largest planets, Jupiter and Saturn, suffered the smallest losses.

3. The geological phase, lasting till the present, during which the planets evolved mostly by their own internal processes and when no major dynamic or physical alterations were imposed either by the Sun or by other planets through perturbations.

It is concluded that the obliquities attained their present magnitudes during phase 2 of this evolution because solar heating and resulting mass loss by the planets could have introduced large increases in the obliquity. On this basis, the obliquities observed for the majority of planets can be understood, but Neptune, with its retrograde rotation, requires a separate explanation. It is possible that the curious obliquities of Uranus and particularly of Neptune result from a close stellar encounter with the solar system during its proto-planet phase or, alternatively, that a peculiar system of turbulent eddies existed in the outer portions of the solar nebula.

While the satellite systems of Uranus and Neptune are opposites, exemplifying perfect order and extreme disorder, respectively, the planets themselves are remarkably similar. This apparently stems from the similarity of the two masses, 14.6 and 17.3 times that of the Earth, respectively. Their similar location near the outer edge of the planetary system is probably a lesser factor. This generalization is based on a planetary survey from which it appears that total mass, rather than distance from the Sun, is the prime factor in determining planetary composition. Dependence on mass is attributed to the evolution of proto-planets, which are assumed to have been essentially of solar or cosmic composition. Thus, in spite of process 2 above, the large proto-planets were able to produce planets of essentially solar composition, whereas small proto-planets lost all the volatile substances and produced cinder-type planets, basically composed of silicates and iron. The terrestrial planets are thus at one end of the scale, the giant planets Jupiter and Saturn, of essentially solar composition, at the other. Uranus and Neptune are intermediate.

Before we consider what has been inferred about the internal structure of Uranus and Neptune, I shall review the direct evidence on the composition of their atmospheres. Spectral analysis of planetary atmospheres is a very potent tool because it leads to conclusions not only about the atmospheric compositions but also about atmospheric temperatures and pressures. The temperature and pressure data are derived from the detailed structure of absorption bands produced by the atmospheric gases observed in the reflected sunlight. The spectra of Uranus and Neptune have an extraordinarily rich structure. Numerous strong absorption-band systems are present which increase in intensity as one moves from the green to the red. In the near-infrared they are so strong that only small islands of reflected sunlight remain. The strong absorptions in the orange and red cause the visual colors of Uranus and Neptune to be peculiar. Uranus has a light-bluish color whereas Neptune is greenish blue. This small color difference is caused by the even stronger orange and red absorptions in the Neptune spectrum as compared to that of Uranus.

The strongest absorption bands in the visible part of the Uranus and Neptune spectra—those responsible for the colors of these planets—are due to methane. They were first observed in visual spectroscopes nearly a century ago by A. Secchi in Italy and W. Huggins in England. The first good photographic records, showing much more detail and extending into the near-infrared, were obtained by V. M. Slipher at the Lowell Observatory in Arizona some sixty years ago. R. Wildt in Germany found in 1931 a few absorptions in the Jupiter spectrum to be due to ammonia and the strong bands common to Jupiter and Uranus to be due to methane. T. Dunham obtained greatly improved spectra of Jupiter, Saturn, and Uranus in 1932–33 with the powerful 100-inch Mt. Wilson telescope and added laboratory comparison spectra with a 40-meter absorption tube. He confirmed Wildt's identification of the visible and near-infrared absorptions in these outer planets. A new era in planetary chemistry was thus inaugurated, with the gradual recognition that the Earth and possibly Mars and Venus have oxidizing atmospheres and the giant planets reducing atmospheres, that is, atmospheres that are hydrogen-rich and thus contain saturated hydrogen compounds, ammonia (NH_3), methane (CH_4), and probably water (H_2O) in the deeper and warmer layers. It became increasingly clear that the contrast between the terrestrial and the giant or Jovian planets is a direct consequence of the planetary composition, which, in turn, is determined by the initial proto-planet mass. A large mass was able to retain even the light gases, hydrogen and helium, which make up 99 percent of the weight of cosmic matter; smaller masses were nearly completely deprived of them, primarily during the very brilliant phase of the Sun immediately following its formation as a star.

The question now arises whether Uranus and Neptune show absorptions in their atmospheres due to constituents other than methane. Ammonia is observed only in Jupiter, apparently because it is frozen out at the very low temperatures prevailing in the atmospheres of Uranus and Neptune. For the same reason, but even more strongly, water vapor is absent in these spectra. One other constituent has been observed in recent years, however—namely, molecular hydrogen, which we would expect to be present in enormous abundance. The first such observations were made by the writer in 1946, and these disclosed a curiously shaped absorption in the near-infrared that in 1952 was attributed to hydrogen gas under high pressure. The normal red and infrared spectra of molecular hydrogen are "forbidden" because of the symmetry of the hydrogen molecule, composed as it is of two identical atoms. Hydrogen under pressure, however, undergoes certain asymmetries that permit the normally forbidden spectrum to appear faintly. The strength of the absorption is a measure of the pressure to which the hydrogen gas is being subjected. These pressure-induced absorptions would be totally invisible were it not for the fact that hydrogen is so extremely abundant in the atmospheres of Uranus and Neptune and, further, that these atmospheres are essentially free of condensation products down to very great depths.

Jupiter and Saturn also contain vast qualities of hydrogen in their atmospheres, but the penetration into these atmospheres is limited by the condensation products of ammonia, which cause impenetrable cloud covers on these planets. The spectra of Jupiter and Saturn are thus due to the sunlight reflected by these cloud covers, with absorptions added as caused by the atmospheric gases above the clouds. Because of the much lower temperatures in the Uranus and Neptune atmospheres, the ammonia cirrus cloud deck will be situated so far down in the atmosphere that visual penetration is no longer limited by the clouds but instead by the molecular Rayleigh scattering in an almost hazefree atmosphere. This situation has some interesting consequences that are troublesome to the spectroscopist: in strong absorption bands we see down to only a comparatively shallow depth, but in weak bands our vision penetrates deeply. The composition of the planetary atmospheres therefore cannot be determined from measures of band strengths and simple comparisons with corresponding laboratory measures, but requires rather complex analysis involving the scattering properties of an immense atmosphere.

Analysis of the pressure-induced absorptions due to hydrogen has given the following approximate model, published in 1952, for the average of the visible Uranus and Neptune atmospheres (which are very similar): 135 kilometer-atmospheres of hydrogen, 370 kilometer-atmospheres of helium, and 3 kilometer-atmospheres of methane. The pressure at the base of this column is about 9 atmospheres (terrestrial). The mean molecular weight is about 3.55 (hydrogen is 2.0, helium 4.0, methane 16). More recent work has not basically altered these conclusions, though the methane content may be slightly higher.

Let me now turn from the atmospheres of Uranus and Neptune to the bulk compositions of these planets. These compositions are, of course, not directly accessible to observation. The procedure rests on the use of measured bulk quantities—namely, planetary mass, diameter, and mean density—to test models based on plausible physical assumptions. Thus one may consider, first, whether the above observations can be satisfied if we assume that the planet is composed of solar material, that is, approximately 80 percent by weight of hydrogen, 19 percent by weight of helium, and a mixture of heavier elements based on analysis of the solar atmosphere. To carry out this examination requires knowledge of the physical properties of hydrogen and helium under the very high pressures that must exist near the centers of these planets. Such theoretical information is now available with reasonable accuracy. It is found that no self-consistent models of Uranus and Neptune can be constructed on the assumption of solar or cosmic composition. In this respect these planets differ drastically from Jupiter and Saturn, where the cosmic composition model appears to be a close approximation.

Next, one may attempt to satisfy the observations by a modified mixture of elements, in which the hydrogen and helium contents are suitably decreased. Such analyses have been carried out during the past decade. Results published in 1965 suggest that Uranus has a rock and metal core contributing about 10 percent of the total planet mass; for Neptune the derived figure is 11 percent. The bulk of the planet mass is found to be in the form of ice, mostly contributed by H_2O. An outer shell of hydrogen, helium, and neon gases, amounting to some 16 percent by weight for Neptune and 20 percent for Uranus, completes the model. These values are still quite provisional but probably correct as to order of magnitude.

• 21

Asteroids, Comets, and Meteors

JOHN A. WOOD

Besides the nine well-known planets that circle the Sun, and their satellites, the solar system contains a great deal of (metaphorical), chaff or debris—vast numbers of small bodies also in orbit about the Sun. These are the asteroids and comets, commonly less than one ten-thousandth as massive as our Moon. Scientific interest in them is quite out of proportion to their size: we may be able to learn more about the origin of the solar system from asteroids and comets than we can from the larger planets.

The asteroids are minuscule planets that orbit in the space between Mars and Jupiter, just at the dividing line in the solar system inside of which the dense, terrestrial planets occur and outside of which are the low-density, gas-rich major planets. About two thousand (1,660, to be more exact) asteroids have been identified, but this is known to be only a small fraction of the total asteroid population. Most of them travel in orbits of low eccentricity and inclination, not very different from the orbits of the larger planets, and they stay outside the orbit of Mars. However, a few travel in more eccentric orbits. Forty-two are known to cross inside Mars' orbit at perihelion, and eight of these even cross the Earth's orbit.

Only four asteroids are large enough to appear as disks of light in a telescope. These are named Ceres, Pallas, Vesta, and Juno. The others are simply points of light. Diameters of the four disks can be measured: they range from 770 kilometers, about the size of Spain or France, down to 193 kilometers. Because we have no knowledge of the masses of these asteroids, their densities cannot be computed. But their surfaces have substantially the same optical properties—reflective power, color, and tendency to scatter and polarize light—as the lunar surface, and so it seems safe to conclude that they are composed of dense rocky material as the terrestrial planets are.

We have no direct knowledge of the sizes of the rest of the asteroids, those that are visible only as points of light. But the brightnesses of many are known; they were measured at the Yerkes and McDonald Observatories by Gerard P. Kuiper and his co-workers, who completed a monumental survey of the asteroids in 1957. The bigger an asteroid is, the brighter it must appear, and so brightnesses can be converted to sizes if the reflective efficiency or albedo of the asteroidal material is known. The albedos of the four largest asteroids have been measured; if we assume the rest of the asteroids to have the same albedo as the average of these, then the following can be said about the total asteroid population: there are about 14 asteroids of diameter larger than 100 kilometers, about 300 asteroids larger than 30 kilometers, and about 2,000 larger than 17 kilometers. The dimmer and smaller asteroids are, the more abundant.

Asteroids much smaller than these cannot be seen in the telescope, but if we may extrapolate from the foregoing relationships, there must be some 3 million asteroids greater than 1 kilometer in dimension, and very approximately a million million objects larger than 7 meters. These colossal numbers make it sound as though the solar system consists mostly of asteroids, but if we sum the masses of all of them, we find that it amounts to only about 3 percent of the mass of the Moon.

Curiously enough, the brightness of many asteroids is not constant but varies periodically. They brighten and dim every few hours. This has been interpreted to mean that they are not spherical in shape, but are irregular oblongs, spinning in space. As a broad face turns into view, the asteroid appears brighter than when we see it end on. When the asteroid Eros passed near the Earth in 1931, its oblong shape could actually be observed. It was about 22 kilometers long and 6 kilometers thick.

These irregular shapes are thought to mean the asteroids are fragments, pieces of some larger planet or planets that have been disrupted, probably by colliding with one another. Certainly, collisions between the present asteroids must often occur. The population distribution of asteroids—the relationship of abundances to dimensions—takes the same form as the comminution law, the mathematical expression that predicts the sizes of particles yielded by industrial rock-crushing machines. The Polish astronomer S. Piotrowski has calculated that the average asteroid suffers a crushing collision every few hundred million years. The asteroid belt acts as a giant grinding mill, perpetually reducing the asteroids to smaller and smaller size. Piotrowski estimates that this process yields several thousand million tons of fine material—dust and sand and pebbles, so to speak—every year.

Not much can be said about how many asteroids were present in the beginning, before the crushing began. There is a break in the population distribution of asteroids, and G. P. Kuiper has suggested that most of the hundred or so larger asteroids that fall on one side of the break may have originally been planets, more or less undamaged, while the abundant smaller asteroids to the other side of the break are collision fragments. Perhaps there were no more than a few hundred asteroids when the solar system was formed.

Comets are about as different from asteroids as they can be. For example, their orbits are not as nearly circular as most asteroids' are. Some, the periodic comets, travel in elongated ellipses, typically reaching perihelion near the orbit of Earth or Mars, then swinging out to aphelion beyond Jupiter. About a hundred of these have been discovered. Others, the near-parabolic comets, fall toward the Sun from vastly distant regions of interstellar space, from as much as a light-year away. Usually they swing around the Sun and then vanish back into the depths of interstellar space, never to return. Several hundred of these have been recorded.

There is another significant difference between comets and asteroids. The great majority of asteroids stay near the plane of the ecliptic and have direct motion, that is, they orbit the Sun in the same direction as the principal planets. Comets, on the other hand, approach the Sun from more or less random directions, both in and out of the plane of the ecliptic, and roughly half of them orbit the Sun in a retrograde direction.

The appearance and behavior of comets and asteroids is quite different. Asteroids are simply small lumps of inert planetary matter, reflecting sunlight. Comets have diffuse heads and long, sometimes strikingly beautiful tails. Both heads and tails are caused by gases and small solid particles, streaming out from the center or nucleus of the comet.

Neither the head nor the tail is present while a comet is far from the Sun. The head, or coma, typically appears after a comet has passed well inside Jupiter's orbit. What we see consists mostly of neutral gas molecules—compounds of carbon, nitrogen, hydrogen and oxygen—moving away from the nucleus at something like one kilometer per second. Sunlight excites these gases and causes them to fluoresce, emitting light at wavelengths characteristic of the molecules present. Spectrographic studies of comet heads and also tails have given us a fragmentary knowledge of the molecules that comprise them. Visible comet heads can be huge, up to a million kilometers in radius on occasion.

Tails usually appear about the time a comet crosses the orbit of Mars. Often two tails are displayed: a gas tail and a dust tail. The gas tail consists of ionized molecules; the spectra observed are again of compounds of carbon, nitrogen, hydrogen, and oxygen. The tail points straight out from the nucleus and away from the Sun, whether the comet is coming or going; it does not trail out behind the comet as it moves through space. Apparently the solar wind, that perpetual flux of protons and other ions that streams outward from the Sun at velocities of roughly a thousand kilometers per second, is responsible for comet tails; it ionizes the gas molecules, and pushes them along, away from the Sun. But just how it does this is poorly understood. A simple collision process between solar ions and cometary molecules has been shown to be inadequate. It seems likely that a coupling of magnetic fields is involved—magnetic fields embedded in the solar wind plasma and in the tail gases. Tail gases stream away at speeds of tens or hundreds of kilometers per second and often stretch to lengths of twenty or thirty million kilometers.

Cometary nuclei shed small solid particles as well as gases, and the pressure of sunlight pushes the smallest of these, the "dust," outward and away from the Sun, forming them into a dust tail. If light pressure were the only force acting on them, they would stream straight out like the gas tail, and would mingle with it. But they are also strongly influenced by gravitational forces, and the interaction of light pressure and gravity gives rise to gracefully curved dust tails. Reflection of sunlight from the dust makes these tails visible; the optical properties of the reflected light allow us to deduce particle sizes of the order of one-thousandth of a millimeter. Fred L. Whipple of the Harvard College Observatory has estimated that comets shed about a thousand million tons of dust into the solar system each year, an amount comparable to Piotrowski's estimate of the asteroidal contribution.

What about the nucleus itself? This is a tiny object, only a few kilometers in dimension. It reflects sunlight and appears as a bright spot in the center of the coma. Whipple has theorized that the nucleus is an icy conglomerate or, more graphically, a "dirty snowball": a mixture of ices of various compounds, probably including water and ammonia, and of dust and larger solid particles of earthy composition. The theory is unproven but seems almost certainly correct; only the sublimation of ices and release of embedded solid matter, as a comet nears the Sun and is warmed by it, seem capable of explaining the steady streaming away of matter. This means, of course, that comets gradually waste away; after a number of passes near the Sun, their ices must be exhausted and their dust dispersed: they cease to exist. The mean lifetime of comets is not known, but an average of the estimates of various writers comes to about a hundred passes.

The origin of comets is a great puzzle. Apparently all of them first come to us in near-parabolic orbits. A few are gravitationally perturbed by passage close to planets, and their orbits are changed into closed ellipses. These are then periodic comets, revisiting the Sun until it destroys them. An object traveling in a parabolic orbit never reaches aphelion: it simply approaches a zero velocity as it recedes to an infinite distance from the Sun. An object held nearly motionless a vast distance from the Sun and then allowed to "drop" toward it would approach the Sun in a parabolic orbit, and apparently this is how the comets come to us. H. J. Oort of the Leiden Observatory in the Netherlands has concluded that the solar system is surrounded by a vast number of cometary ice masses, 100 billion of them. These form a huge cloud, roughly a light-year in radius, about the Sun. Ordinarily comets within the cloud are moving very slowly, or hardly at all, relative to the Sun; but from time to time some of them are gravitationally perturbed by neighboring stars, and of these a fraction are given pushes toward the inner solar system, enough to cause them to fall around the Sun in near-parabolic orbits.

Oort's model raises as many questions as it answers. How did this cloud of comets come into being? Apparently it is not fed from interstellar space or from other stars, for then the comets would not be hanging nearly motionless in it. They would have substantial velocities relative to the Sun, and those that fell around the Sun would travel in hyperbolic, not parabolic, orbits. Hyperbolic orbits are not observed, other than those accounted for by planetary perturbations. It appears that comets are truly creatures of the solar system, in spite of their present remoteness from it. G. P. Kuiper in the United States and B. Y. Levin of the Soviet Union believe that the comets condensed in the outer reaches of solar system at the time when the planets themselves were forming. Subsequently, perturbations by Neptune hurled them away from the solar system. Those that were thrown at near-parabolic velocities ended up in Oort's comet cloud.

We have seen that decaying comets and colliding asteroids are perpetually adding dust to the solar system. It might seem that the plane of the ecliptic would become clogged with dust. In fact there is a dust cloud, and sunlight reflected from it causes a zone of very faint illumination that follows the ecliptic plane across the night sky. This is called the zodiacal light. But there are processes that remove dust from space as well: solar light pressure "blows" the smallest particles (less than about one thousandth of a millimeter) away from the solar system; larger fragments, up to a few centimeters, are slowed in their orbits and sent spiraling into the Sun by the Poynting-Robertson effect;[1] and the planets tend to capture particles of all sizes. So it is likely that the zodiacal cloud has reached a steady state by now, with dust being removed from it about as fast as it is added.

---

[1]Poynting-Robertson effect: small orbiting bodies absorb and re-emit solar energy. Re-emission is in all directions, but radiation sent out ahead of a moving body is crowded together, its wavelength slightly shortened. Conversely, the wavelength of radiation sent backward is stretched out. These wavelength shifts have the effect of exerting a slight drag on small orbiting bodies and slowing their motion.

If dust particles and small fragments are abundant in the inner solar system, then the Earth must continually be colliding with them and capturing them. The collision velocity is always at least 11 kilometers per second, for the gravitational attraction of the Earth alone guarantees this much. If the particles have been traveling in elongated cometary or asteroidal orbits, encounter will be at 20 to 30 kilometers per second. Particles entering the Earth's atmosphere at such high velocities are heated to incandescence by friction with the air, and in some cases they are melted or vaporized. When this happens we see a momentary streak of light in the upper atmosphere: a meteor.

There are shower meteors, which fall at particular times of the year; examples are the Perseid meteors, about August 12, and the Geminids, about December 12. There are also sporadic meteors that appear all during the year. Meteors can be observed with telescopes or by radar. Careful studies of meteor trails and the rate at which the atmosphere is able to decelerate the meteoroids have shown that two fundamentally different types of objects are entering the upper atmosphere.

One type has the strength and mass density of normal planetary matter. These objects are abundant among the very bright, large meteors and fireballs. They fall sporadically, and almost certainly are asteroidal fragments.

The other type is fragile and has a low net mass density, less than that of water. Apparently these particles consist of porous, skeletal material. They give rise to almost all the faint meteors, high in the atmosphere. All the shower meteors and most sporadic meteors are caused by this type of material. It is certainly cometary in origin. Many of the periods of shower activity occur when the Earth passes through the orbital paths of known comets.

Very large meteoroids, hundreds of grams or more, are able to make their way through the atmosphere without being totally destroyed. Air friction melts their surfaces, but little heat has a chance to penetrate to their interiors. Almost two thousand of these have been collected on the Earth's surface and saved over the years: they are the meteorites. All are of dense rocky or metallic material, and they fall sporadically. None of the low-density cometary material has made its way down and been recovered. Thus the meteorites are probably pieces of asteroids, although Harold C. Urey of the University of California has suggested that some may have been derived from the surface of the Moon, thrown out by impacting asteroids or comets.

At present meteorites are the only samples of extraterrestrial material that we can get our hands on, to analyze and study by all the techniques of chemistry and mineralogy. In later years we will have samples of the lunar and possibly the Martian surface, but for the time being meteorites occupy a unique position. This is ample justification for the programs of meteorite study that are presently being conducted, but there is an even better reason. Radioactive dating has shown that the meteorites are very old, in the sense that they have been preserved essentially unchanged for a very long time—about 4,600 million years. No terrestrial rocks have survived change this long because rocks in the Earth's crust are continually being destroyed and reconstituted by various processes, especially melting and weathering. But meteorites seem still to have characteristics that were impressed on them in the very earliest years of the solar system.

The iron meteorites and some stones are igneous in character, and tell us of a time of high temperatures and melting that affected the newborn planets. The chondritic stony meteorites, on the other hand, seem to be undifferentiated and possibly even primordial planetary matter—still in the same state the planets assumed when they first accreted. It appears that a great deal can be learned about the processes that formed the terrestrial planets and shaped their early history from studies of meteorites.

Let me now return to the asteroids and comets and ask how fruitful it might be to use rocket-borne space probes to approach them and to study them at close range. We can contemplate experiments at three levels. First, a spacecraft could be made to pass very close to an asteroid or comet, to take a close look at it. Second, a spacecraft might be made to land on the object, attach itself to it, perform analyses of various sorts, and radio the data back to Earth. Finally, we might consider a spacecraft that would land, take a sample, and then bring it back to Earth.

The fly-by experiment would not be difficult; it should be no harder than the Mariner IV mission that sent back pictures of the Martian surface. The other two experiments are obviously much harder, not only because of the delicacy of the landing maneuver but because asteroidal and cometary orbits are so different from the Earth's. A very radical change has to be made in the spacecraft's velocity vector after it is launched if it is to approach an asteroid or comet slowly enough for a safe landing. This means that a great deal of thrust has to be provided, and so a powerful rocket system would be required. However, such a mission appears to be marginally possible if a particularly favorable asteroid or comet is chosen and if a large booster rocket of the Saturn type can be allocated to it.

My own view is that a great deal more could be learned from comets than from asteroids by means of space experiments. In part this is because we are so ignorant of comets: there are more questions and puzzles about them than there are known facts, whereas asteroids seem to be fairly straightforward fragments of orbiting planetary matter. Also, since the meteorites have already been studied in detail, and samples of the lunar surface (at this writing) soon will be, it is probably true that the cream will already have been skimmed from the data that an asteroid probe would return. A sample returned at great labor and expense from an asteroid would in all likelihood turn out to be very similar to the meteorites in our collections. A comet sample, on the other hand, is certain to be quite different from anything we have ever seen before.

The prospect is indeed intriguing that we might one day be able to study material that has spent most of its life a light-year away from the Earth. It has been suggested that comets are samples of primordial material of the type that went to make up the major planets, Jupiter, Saturn, Uranus, and Neptune: this would make them counterparts of the chondritic meteorites, which appear to be primordial samples of the terrestrial planets. Both kinds of primordial material should have a fascinating story to tell.

IV

THE COSMIC ENVIRONMENT

• 22

The Optical Universe

ARTHUR D. CODE

The ultimate task of the astronomer is to describe the nature of the physical universe—that is, to determine the present distribution and state of matter and energy in the universe and to inquire into its origin and evolution. The partial picture that is revealed by the light received in the visible region of the spectrum is what we call here the *optical universe.* The studies of the optical universe extend back to the earliest scientific inquiries of man.

It is only recently that instruments have been devised to extend man's senses beyond the limitations of his eyes, into the long-wavelength region of the radio spectrum and the short-wave X-rays. Our present knowledge of the physical universe results from a synthesis of the information provided by all these studies. In this discussion we shall be primarily concerned with the information derived from visible light—namely, the radiation that we can see. As a useful starting point let us consider a question frequently asked of an astronomer: How far can we see?

On a clear night you can easily see the great spiral nebula in Andromeda. This object, a galaxy similar to our own Milky Way, is some 20 million million million kilometers away. The light that falls upon our eyes today started its journey through space about

2.25 million years ago. Among the faintest images on photographic plates taken with the large reflecting telescopes are objects some thousand times more distant still, and it is believed that many of the sources measured with radio telescopes extend man's view even further out into space and back in time. In years past, when the skies were darker and clearer than they are today in our industrialized civilization, the ancient Greeks also looked at Andromeda, but had no way of appreciating the extent of their vision. They lacked the tools required to extend their senses and, more important, almost the entire body of physical theory which has been developed only in the last 350 years. Indeed even Kepler, whose laws of planetary motion finally displaced those of the ancient Greeks, believed that all the stars were contained in a not too distant sphere only a few kilometers thick. Nor did he realize that the Sun was a hot gaseous sphere some hundred times the diameter of the Earth and the stars were other distant suns, many smaller and some much larger than our own Sun.

How man has extended his investigations further and further beyond the boundaries of his tiny world is a story of advancing technology and bold new ideas. In one sense, however, it is a very simple story, for with the exception of the exciting prospects of exploring the solar system with space vehicles, the astronomer is restricted to observing the radiation received from celestial objects. He cannot perform experiments in the same way as physicists or chemists. The analysis of this light or electromagnetic radiation consists of determining the direction of propagation, the intensity, the spectral distribution, and the state of polarization of this radiation as functions of time. All of observational astronomy may be described in terms of one or more of these types of measurements.

The astronomical telescope is designed to collect light and to increase the precision with which the direction from which it is coming can be determined. The larger the diameter of the telescope, the more radiation can be collected, which in turn means that fainter objects may be observed or more detailed analysis can be carried out on the brighter objects. Auxiliary instruments such as photographic plates, spectrographs, and photocells make it possible to increase the precision of the measurements and extend them to the ultraviolet and infrared regions of the electromagnetic spectrum, where the eye is not sensitive. The conceptual tools of the astronomer are the basic laws of physics determined by experimentation here on the surface of the Earth. It is assumed that the physical theory appropriate to terrestrial experience may be applied elsewhere in the universe, an assumption that is then checked whenever possible. Our understanding of the nature of the physical universe is therefore based upon the interpretation by theory of the radiation received from celestial objects.

How do we determine the distances to stars and to galaxies? The distance to the nearer stars is found by accurate measurements of the direction from which their light arrives. This measurement, called a trigonometric parallax, is similar to the technique often employed by surveyors in mapping parts of the Earth. If many photographs are taken of a star field over the course of a few years, the nearer stars appear to move back and forth with a period of one year due to the orbital motion of the Earth about the Sun. Indeed, the motion of the star is just that which would be displayed by the Earth as seen from the distant star. This angular displacement is small but measurable for the closest stars and undetectable for the more distant objects. The direct trigonometric method fails when the distance exceeds about 100 light-years.

One way of extending our measurements to greater distances depends on measuring the intensity of stellar radiation. We know that as the distance of a light source is increased its brightness decreases. In fact, its brightness varies inversely as the square of its distance. Thus if two stars were of the same intrinsic luminosity but one was ten times as far away, this more distant star would be one-hundredth as bright. If all stars were of the same intrinsic luminosity, we could determine stellar distances simply from their apparent brightness. Near the close of the eighteenth century William Herschel applied this technique by assuming that all stars are similar in brightness to the bright star Sirius. Within the framework of this approximation, he was able to determine the distances of stars in units of the distance to Sirius and found that the stars formed a flattened system concentrated to the Milky Way, and that the Sun was located some distance from the center. This picture is basically correct although, as Herschel himself must have realized, the assumption that all stars are of the same luminosity is very far from the truth. Some stars emit a million times as much light as the Sun while other small cool stars barely radiate at all. This being the case, how are we to apply this brightness technique of measuring distances?

There must be a method of independently determining the intrinsic luminosity of a star. Some stars show periodic variations in brightness. From their light curves, or variation of brightness with time, it has been found that some of these variables are double-star systems, that is, two stars revolving about each other. As they revolve, their total light is periodically reduced when one eclipses the other. Not all variable stars are double systems, however. Some are found to be single stars which pulsate, and as they expand and contract their brightness varies in a characteristic way. It has been established that pulsating variables with the same period of light variation and the same shape of light curve have very nearly the same intrinsic luminosities. Thus if we determine the period of these stars, called cepheid variables, we can find their distance by comparing their apparent brightness with their intrinsic brightness. The cepheid variables are one of the principal distance indicators employed for the study of the nearer extragalactic nebulae, such as the Andromeda galaxy. The determination of the dimensions of our own Galaxy or Milky Way depends in part on the use of pulsating variables, and some forty years ago such measurements set the size of our Galaxy at about ten times the dimensions we now believe to be correct. The reason for this gross error is the fact that interstellar space is not empty but contains gas and small dust grains that dim and redden the star light on its passage from the star to us. A star may not be faint, then, because it is at a great distance, but rather because it is behind an obscuring cloud of dust.

Quite simply then, stars do not all look the same brightness to us because (1) they are at different distances, (2) they are of different sizes, (3) they have different surface temperatures or surface brightness, and (4) they are obscured by different amounts of interstellar matter. A large part of stellar astronomy is concerned with separating these different reasons for the variation in apparent brightness of the stars. If we are to better answer the question "How far can we see?", we must therefore know more about the nature of stars and of interstellar matter.

One of the most important tools of the astronomer has been the spectrograph. A spectrograph employs a prism or grating to disperse the light of different colors or wavelengths and record the spectrum on a photographic plate. Thus, instead of a single star image, the star light is spread out across the plate so that the amount of radiation at each wavelength, within the sensitivity range of the photographic emulsion, may be measured. Although in the radio region the wavelengths are commonly measured in meters or centimeters, the wavelengths of visual light are usually expressed in terms of angstroms, a unit of length equal to one hundred-millionth of a centimeter, or about the size of a single atom. The eye, for example, has its maximum sensitivity in the green near 5,500 angstroms and extends to the violet to about 4,100 angstroms and to the red to about 7,500 angstroms. Photographic plates are available that are sensitive from the X-ray region on into the infrared around 11,000 angstroms, although the opacity of the Earth's atmosphere prevents radiation short of 3,000 angstroms in the ultraviolet and much of the infrared spectrum from reaching the surface of the Earth.

If in the laboratory we were to illuminate a spectrograph with an incandescent lamp, like an ordinary light bulb, we would find that radiation of all wavelengths was present in the focal plane of the spectrograph; we would observe a continuous spectrum. If we made the lamp hotter, not only would it appear brighter overall, but in particular the blue end of the spectrum would become brighter relative to the red. If, on the other hand, we were to use a neon lamp or a fluorescent lamp to illuminate the spectrograph, we would find radiation at discrete wavelengths. We would see what is called an emission line spectrum. The particular spectral lines or wavelengths that would be bright are characteristic of the chemical element giving off the light. If we were to replace the neon by helium, a different bright line spectrum would appear, and indeed a chemical analysis of an unknown sample can be made by examining the line spectrum produced when a sample is heated. Now, laboratory studies show that the intensity of a particular spectral line depends upon the abundance of the element responsible for that line and upon its temperature and pressure. It is thus possible from spectral analysis to determine not only what elements are present but in what quantities, and what the temperature and pressure of the gas are.

Finally, if we were to shine the light from our incandescent filament lamp through the neon lamp and examine the spectrum, we would find that if the neon lamp was more intense than the incandescent lamp, the spectrum would consist of a continuous spectrum with superimposed emission lines characteristic of neon. On the other hand, if the incandescent lamp was brighter, the continuous spectrum would show gaps, or absorption lines, at the neon wavelengths. All these physical processes are well understood, and it is possible to predict from theory just what the spectrum should look like even for conditions that have not been measured in the laboratory.

When we place a spectrograph at the focus of a telescope and examine the spectrum of a bright cloud of interstellar gas like the Orion Nebula, we see an emission line spectrum. It is found that the most abundant atom is hydrogen, the next most abundant element is helium, with lesser amounts of the heavier atoms. The temperatures of these bright diffuse nebulae, which incidentally are heated up by a nearby hot star, are usually about 10,000 degrees Celsius. The pressures correspond to that produced by about ten to one thousand atoms per cubic centimeter.

If we focus our telescope on a star, the spectrum is usually found to be a continuous spectrum with absorption lines due to the same elements found in the interstellar clouds. A simplified explanation of the observed spectrum is that the continuous spectrum comes from the lower, hotter opaque layers while the cooler outer layers of the star absorb radiation at the characteristic wavelengths associated with their chemical composition, temperature, and pressure.

Theory would indicate that if two stars have identical spectra they should be identical in all other respects, too. We can check this suggestion by examining the spectra of the nearby stars whose distance can be determined by trigonometric parallaxes. It is indeed found that stars with the same spectra have similar luminosities. We may therefore determine the distance of a star beyond the range of our trigonometric technique by measuring its apparent brightness and obtaining a spectrum of the star. From the spectrum we may deduce its intrinsic luminosity and then, from the apparent brightness, its distance. From such measurements it is even possible to correct for the effect of the interstellar dust. Not only does the dust dim the star light but it makes the star redder, just as the dust in the earth's atmosphere reddens the Sun near sunset. From the star's spectrum, then, we know not only the luminosity of the star but also what its color should be. If it is redder than nearby stars with similar spectra, we can deduce just how much the light has been reduced by the dust and make the appropriate correction to our distance determination. There are several ways of checking the accuracy of our photometric distances. Stars often occur in groups or clusters. The members of a single cluster were in general formed at the same time, of the same material, and are at essentially the same distance from us. The stars belonging to the cluster differ in intrinsic luminosity, spectra, and other properties because they are of different mass, the most massive being the most luminous. If our technique of determining photometric distances is a good one, we should get the same distance for each of the cluster members; and indeed we do.

There are two distinct types of star clusters characterized by their physical appearance, their spatial distribution, and their motions in our Galaxy. The Galactic clusters are irregular in shape, are confined close to the plane of the Milky Way, and move slowly with respect to the Sun. The globular clusters are compact and spherical in shape, are loosely distributed about the center of our galaxy, and have high velocities. It was found that, while the stars in galactic clusters showed the same relation between luminosity and spectrum as the nearby stars, this was not true of the stars in globular clusters. A more detailed study of the spectra of globular cluster type stars revealed that the spectra were, in fact, not the same as those in galactic clusters. The globular cluster stars had much lower abundances of the elements heavier than helium than did the Sun or galactic cluster stars. This and other related facts have led us to a picture not only of the formation of the stars but of the formation of the chemical elements themselves.

Studies of the distance and motion of stars have shown that we live in a large stellar system containing some 100 billion stars concentrated in a flattened disk. The Sun is located near the edge of the disk, rotating about the center in about 200 million years. The star density increases toward the center, forming a central bulge. The brighter stars and interstellar gas and dust form spiral arms in the disk. Superimposed on this system is a spherical distribution of stars and globular clusters which have low heavy-element abundance and, because they are not revolving with the disk as is the Sun, their velocities appear to us to be large. This whole system is called a spiral galaxy. The Andromeda Nebula is also such a spiral galaxy and is one of the Milky Way's nearest neighbors.

I have been primarily concerned about determining distances because, until we know how far away an object is, we cannot really know just what the object is. Let us now assume that we have collected extensive data on the distances, luminosities, and spectra of many stars. Having collected these vital statistics, can we understand the structure of stars? This is a task for the theoretician. He attempts to construct on paper a model star of a given mass, in which the outward pressure forces tending to blow up the star are just balanced by the inward pull of gravity which tends to compress the star. The pressure and therefore the temperature will be greatest in the center and decrease outward. This means that energy will flow outward and finally escape as light. The test of his model star is: Does the light that escapes have exactly the same intensity, spectral distribution, and spectral line strengths as that of a real star? If he found that the sodium lines in his model were too weak, for example, he might add a bit more sodium until the agreement between theory and observation was satisfactory. This model star then describes in detail the structure, physical conditions, and chemical composition of a real star.

As the light moves outward through the star and escapes into space, the star loses energy. If there were no source of energy in the center of the star, it would slowly contract until it settled into a dense, cool mass or violently exploded. Stars of mass similar to the Sun do end their existence this way, eventually becoming what we call white dwarfs. More massive stars shed some of their mass back into interstellar space. But if gravitational contraction were the only source of energy available to a star, the Sun would have shone for only a few million years. We know, however, of fossils more than 500 million years old, and geological evidence indicates that the Sun has radiated at essentially its present rate for a few thousand million years; thus some internal energy source is required to explain the luminosity of the Sun and stars. This energy source is thermonuclear reactions.

According to the star model calculations, the temperatures and densities prevailing in the center of stars are such that nuclear processes should go on. What happens is that four hydrogen nuclei are converted to one helium nucleus, liberating the extra binding energy required to hold four separate nuclei together rather than one large one. Since hydrogen is the most abundant element, most of a star's existence is spent burning its hydrogen fuel and radiating this energy into space. Detailed calculations of the thermonuclear-energy generation of stars account for the observed luminosities over the required lifetimes very nicely.

From observations and theoretical calculations similar to those described here, astronomers have constructed the following picture of the life history of the stars in our Galaxy. About 10 billion years ago our Galaxy consisted of clouds of hydrogen or possibly hydrogen and helium. These clouds fragmented and contracted into much smaller clouds, or proto-stars, of sufficient density for their self-gravitation to hold them together. The proto-stars continued to contract, thereby heating up and becoming self-luminous. In this early stage the energy source for these new stars was simply gravitational contraction. Finally, the temperature and pressure in the interior became sufficiently high that thermonuclear-energy generation started. During this period the star stopped contracting and remained at essentially its same radius and luminosity, while the hydrogen in the interior was converted to helium. When the hydrogen core was exhausted, gravitational contraction set in again, raising the temperature in the core until conversion of helium to carbon occurred. In this way, heavier and heavier elements were built up from lighter ones. The massive stars then shed some of their matter back into interstellar space, thus enriching the interstellar medium with heavy elements. New stars were created of this enriched mixture, while dust grains formed in the interstellar clouds.

This process continued through several generations. The youngest stars which spin around the Galaxy in the galactic plane have higher heavy-element abundances than the old remnants of the early generations still found in globular clusters and far from their place of birth.

I have described here techniques and ideas that are reasonably well understood. That is not to say that there is not much work yet to be done in these fields. There are still many puzzles and exciting things left to be discovered about the process of star formation. This is, however, a story of the past. What new challenges lie ahead? One most certainly is the understanding of the formation of the galaxies themselves and the origin and structure of the universe. Some progress has already been made in this direction.

Beyond our own Galaxy, beyond the Andromeda galaxy, are as

many more galactic systems as there are stars in our Galaxy. The fainter and more distant these objects are, the faster they seem to be moving away from us. This fact has led to various cosmological models representing the universe as an expanding system. Recently we have come to realize that intergalactic space contains more than these galactic star systems. There exist faint wisps of gas, exploding galaxies, and fantastic globs of matter apparently radiating energy more violently than anything astronomers have yet observed. Before we can understand these "quasi-stars," as they are called, we must know their distances, a subject of great debate currently.

Many of the new and exciting problems confronting astronomers today have come about as a result of radio astronomy, which has provided man with a view of the universe in the long-wave region of the electromagnetic spectrum. Certainly the ability to get above the Earth's absorbing atmosphere with satellites will also provide new surprises and rich intellectual rewards. Far above the Earth we can explore the universe in the infrared and look for the thermal radiation of proto-stars and proto-galaxies. In the ultraviolet and X-ray region early space experiments have revealed new unexplained objects. Perhaps in the X-ray region, the high-energy end of the spectrum, we may find the source of cosmic rays and link the structure of the universe to fundamental particles that are the building blocks of nature. Nor must we forget the large, high-speed, digital computers that make what were once impossible calculations easy; nor the direct exploration of the planets, which is soon to be a reality and with it the possibility of further unraveling the nature of life and man's small place in the universe.

How far can we see? Perhaps in time to the boundaries of the universe—or to the beginning of time.

• 23

The Radio Universe

JOHN W. FINDLAY

The universe in which we live presents to us appearances differing from one another as we choose different methods of astronomical observation. For very many years astronomers used their eyes, aided by telescopes, for their observations. They increased the sensitivity of these observations as telescopes became larger and as photographic plates, photo-multipliers, and other image-capturing devices improved.

However, although some extensions were made outside the range of visible light into the nearby ultraviolet and infrared regions, the atmosphere of the Earth prevented any considerable excursions to other wavelength ranges. Nevertheless, a very fascinating picture of the optical universe has been derived from these studies.

We live on a planet which is one of several in orbit around a rather ordinary type of star. Our star belongs to our Galaxy, made of stars, gas, and dust. By optical studies, many other galaxies beyond our own had been found even before our own had been mapped in much detail. Studies of the spectra of distant galaxies and of the famous red shift of spectral lines were combined by optical astronomers with other distance indicators to give the first measures of the size of the universe. Such observations were joined with cosmological theories to attempt to understand the physical laws which apply to such enormous extents of space and time.

A very satisfying picture of the optical universe has emerged, but it is still, of course, one that demands answers to very many questions. However, with the growth of radio astronomy, we have had the opportunity to re-examine our universe by observing the radio waves emitted by objects within it. We get a somewhat different picture of the universe. For example, on a night with no moon the bright stars stand out as the most obvious visible objects in the sky. But with radio telescopes, we see almost no stars, yet our own Galaxy shows its presence by radio waves from the tenuous gas between the stars. One of the best-known and most valuable results from radio astronomy has been this ability to study the hydrogen gas within our Galaxy, and to map its density and its motion. The spiral nature of our Galaxy has been confirmed and the complex structure at the galactic center, invisible optically due to absorption, has been observed in detail.

Within our Galaxy also we find bright radio sources, some almost invisible optically, which we believe to be the remnants of supernova explosions. When we look at distant galaxies, we find some whose behavior at radio wavelengths is like our own, but others are very much brighter. We find the sky full of radio sources, many yet not identified with obvious optical objects. Some of these radio sources have pointed the way to the discovery of quasi-stellar sources, a discovery which appears to have added a new and curious type of astronomical object to our picture of the universe.

There are not, of course, great differences between the optical and radio universes: they are complementary pictures of the same thing. The advances and discoveries come in the areas where the pictures differ, but each advance must also make the total picture more consistent and complete.

Unlike optical astronomy, the science of radio astronomy is new because radio itself is relatively new. Although the mathematical theory of the propagation of radio waves was laid by James Clerk Maxwell as early as 1873 and although there were experiments by Heinrich Hertz in radio communication inside a laboratory in 1887, it was not until the beginning of the twentieth century that radio developed as a means of communication. Even in those early days some scientists wondered whether astronomical objects might not emit the same kind of radio waves that were being used for communication. In fact, at that time enough was known about the laws of physics (for example, Planck's law for radiation from hot bodies) to allow us to calculate how much radio energy might be emitted by the Sun. The calculations would have led to a minimum value. So far as I know, no such calculations were made, but there is a record that Edison as early as 1890 planned an experiment to detect radio waves coming from the Sun and Sir Oliver Lodge in 1894 said in a lecture that he himself had attempted to detect rather long-wavelength radio radiation from the Sun.

It was not until the early 1930's that for the first time a definite measurement was made of radio signals coming from outside the Earth. A young engineer, Karl Jansky at the Bell Telephone Laboratories in the United States, was studying for practical reasons what the basic noise levels would be in a projected transatlantic radio telephone system. The performance of such a system, of course, would depend both on the transmitted power used and on the noise level that would exist in the receiver, and it was already known at that time that there was a lot of radio noise generated within the Earth's atmosphere by such things as lightning discharges. To study these, Jansky built an antenna which could rotate so that he could steer the beam, or the direction of maximum sensitivity of the antenna, in any direction. He connected it to a receiver of good quality and measured the noise level that he received. Much of this noise came from the lightning flashes that he was expecting, but he also recognized a basic continuous noise level in his receiver. He identified this as a hissing noise.

What it meant, in fact, was that radio waves in which the electric field was varying in a thoroughly random manner were reaching his antenna. Jansky observed that the level of this signal varied throughout the day, and day by day, and he found that it was strongest when his antenna was pointed in a particular direction in the sky. After a period of some months he realized that the direction from which the radio noise was coming coincided fairly accurately with the part of the sky lying in the direction of the center of our own Galaxy. Jansky had detected radio noise generated within the great disk of stars, dust, and gas which constitute our Galaxy, and he had discovered that the central region of this

Galaxy was a strong emitter of radio waves. Although this very basic discovery in a new science was well publicized and very well described by Jansky in two technical papers, the results did not get the scientific recognition they deserved.

As late as 1937, when I myself started as a young graduate student to do research in ionospheric physics, I remember reading Jansky's papers because they were in a field which was obviously connected with the one in which I was going to work. I can still remember the feeling I had that here was a new and unexplained phenomenon, and I can also remember well the feeling I had that I had already selected the field of research in which I was going to work and could not be directed into a novel area.

In fact, it was not until just before World War II that interest in the new subject of radio astronomy was reopened—and reopened by a young isolated investigator, Grote Reber in the United States. Reber was a radio engineer and a radio amateur. He had read Jansky's work and decided that he could build in his own backyard in Wheaton, Illinois, a radio telescope to test Jansky's results and to search for other radio radiation from the sky. That he did so is now ancient history; his reports were published shortly after the end of the war.

World War II itself might also be said to have rediscovered the subject. Radar receivers in Europe suffered interference which was at first thought to be deliberately manmade to spoil the performance of the radar equipment. But when the interference was studied by scientists like J. S. Hey, it was found to have originated from the Sun. The Sun was very active at that time, and we now know that an active, disturbed Sun is a powerful source of radio waves. So at the end of the war many scientists came back to their universities ready to use the great electronic advances which had been made and with a new scientific subject to explore. Groups formed in England, in Australia, in the United States, and shortly afterwards in several other countries.

The first discrete sources of radio waves in the sky were located. At first these were called radio stars, but it was soon found this was not a satisfactory name since most of the bright stars are not detectable as radio emitters. Most of the new radio sources which were found in the sky were for some time unidentified. As the accuracies of locating these sources became better, one by one they became identified with optical objects. John Bolton in Australia made one of the first identifications when he associated a strong radio source with the Crab Nebula. The Crab Nebula, it will be remembered, has been known for some time as the remnants of a supernova which was observed by the Chinese in A.D. 1054. Soon afterward, a very peculiar pair of galaxies in the direction of the constellation of Cygnus was identified with one of the most powerful radio sources observed on Earth. In the last fifteen years, the identification of radio sources with optical objects has proceeded rapidly. Now we know that the Sun, the Moon, and some of the planets are radio sources. We know within our own Galaxy that clouds of excited hydrogen gas are radio sources; so also are remnants of many supernovae. The neutral hydrogen in our Galaxy emits radio waves, and beyond our Galaxy towards the farthest edge of the universe we detect many other galaxies as sources of radio waves. We also detect and measure the puzzling quasi-stellar sources.

The field of study is great both in the extent and variety of the astronomical objects which can be detected and also in the very wide spread of wavelengths over which radio astronomy can range. Even from the surface of the Earth, where we are shielded at the long-wave end of the radio spectrum by our own ionosphere and at the short-wave end by our atmosphere, we can still make good observations over ranges of wavelengths from perhaps 1 millimeter at the short end to 30 meters at the long wavelength end.

This range of 30,000 to 1 in wavelengths is one of the advantages that radio astronomy has over optical astronomy, where the range available is only about 3 to 1. Of course, optical astronomers are now busy extending their observational range by beginning to put telescopes into space vehicles above the atmosphere. Radio astronomers will soon do the same, but at present the bulk of their observations are still made from ground-based radio telescopes.

Such telescopes have been built in a very wide variety of forms but they are basically similar instruments. Each consists of a highly sensitive, highly stable radio receiver connected to an antenna system arranged in such a way that the direction of maximum sensitivity of the antenna can be pointed to different parts of the sky. The radio receiver must be capable of measuring very weak signals. As an illustration it has been said that the total energy so far collected by radio astronomers in the last twenty years is only the equivalent of the energy available when a snowflake falls to the ground. Thus the radio telescopes have very large collecting areas. The one in Arecibo, Puerto Rico, is a spherical reflecting surface, 1,000 feet in diameter, covering an area of more than 18 acres. This is the largest reflecting radio telescope in the world. Even with such an instrument directed at quite a strong radio source, the available power to be measured is only about 10^{-14} watt. And even the best radio receivers generate within themselves electrical noise which is indistinguishable in character from the radio noise signals collected by the antenna; it is not much less in power than the 10^{-14} watt collected by a large telescope observing a strong radio source. Yet it is possible by comparison techniques, where the signal from the antenna is compared with the signal from a steady source of radio noise, to measure power levels hundreds or thousands of times smaller than the receiver noise itself. Thus very faint radio sources can be detected and measured.

Even such large reflector instruments as the 1,000-foot Arecibo dish still do not permit the location of the radio source in the sky to a high degree of accuracy. The cone of sensitivity or the beam width of such a telescope is simply related to the dimensions of the telescope aperture measured in terms of the radio wavelengths that is being used. Thus the 1,000-foot dish at a wavelength of 75 centimeters has a beam width of about 10 minutes of arc. This accuracy in angular position is very poor when compared to the positional accuracy of optical instruments, and had not methods been developed for improving the positional accuracy of radio telescopes the science would not have developed very far.

A variety of ways has been used for getting better angular resolution. It is possible, of course, to work at shorter wavelengths, but there the requirements for a very smooth reflector surface (again measured in terms of the wavelength) make it difficult to improve the beam width very much. Such instruments as the 140-foot telescope at Green Bank, West Virginia, and the 120-foot telescope of the Lincoln Laboratory in Massachusetts both can work to wavelengths of a few centimeters and have beam widths of 2 or 3 minutes of arc. Where still greater angular resolution is required, we can substitute for the large, single, circular reflector an array of smaller telescopes or we can change the shape of the reflector surface. A telescope recently built in Australia near Canberra is in the form of two perpendicular lines of antennas, each a mile long, one running north-south, the other east-west. By correctly combining the signals received on these two lines of antennas, the resolution equivalent to an aperture a mile wide is achieved, and by electronic means the beam of the telescope can be pointed at different parts of the sky.

An even more elegant way of achieving high resolution has been developed by the group working at Cambridge, England, and is now used at several observatories throughout the world. This technique, called "aperture synthesis," allows the experimenter to make observations to resolutions of a few seconds of arc. It can only be used when the radio source being studied is known not to vary with time, but this is true of most radio sources. Observations are made with a pair of or several small antennas which are widely separated one from another on the ground. The separation of the antennas is changed from one experiment to another either by moving them or by using the fact that as the Earth rotates their apparent separation as viewed from the radio source changes. Successive observations are combined in a computer, and the final result is the equivalent of what would have been obtained from a single telescope as big as the largest separation of the individual elements. With such telescopes, details within radio sources of a few seconds of arc have already been resolved, and accurate positions and sizes have been found for many sources of small angular diameter.

The discovery and study of the quasi-stellar sources is an excellent example of the interdependence of radio and optical astronomy. A few years ago measurements of the positions of radio sources (by the workers at the California Institute of Technology) had become so accurate that, when the optical photographs were studied, the area within which the radio source was known to be contained very few optical objects. In one such case, a faint star-like optical object was the only possible identification to choose for the radio source. At first it was suspected that this might be a star (although, as we have seen, very few stars are radio sources), but optical spectral observations showed a considerable red shift; this, in turn, was most simply interpreted as placing the object at a considerable distance outside our own Galaxy.

Soon many such objects were found, by the same method. The search starts with the radio knowledge that the angular size of the radio source is small. An accurate radio position leads to an optical identification and this in turn to a measured red shift.

The quasi-stellar sources show other curious properties. Some show fluctuations in the light they emit. Some show a similar type of variability in their radio emission, particularly at wavelengths in the centimeter range. About 100 such sources are now fairly certainly known, and the observed red shifts are greater than ever before measured for any objects. Cases exist where the red shifted wavelength is more than three times the emitted wavelength.

One of the most interesting techniques in radio astronomy which is being used to find possible quasi-stellar sources is the application of lunar occultations. As the Moon goes around the Earth, it may pass between an observer on the Earth and the radio source which he wishes to study. As a particular radio source disappears behind the Moon and then as it reappears, the radio signal from the source is cut off and returns. The disappearance and reappearance of the signal is also accompanied by diffraction effects. Although these effects somewhat complicate the picture, they actually add knowledge of the angular size of the radio source itself: the times of disappearance and reappearance can locate the true position of the source in the sky to a few tenths of a second of arc. Cyril Hazard uses the large Arecibo telescope to find possible quasi-stellar sources occulted by the Moon. First, the sky is observed along the track which the Moon is known to follow. A catalogue is made of the faint sources which can be seen; these are generally so faint that they are not known or catalogued by others. Thus approximate positions for those weak sources to be occulted are found. The dates and times of the expected occultations are predicted by the Nautical Almanac Office and the occultations are observed. Thus the precise positions of the sources are found and estimates made of their angular diameters. The Sky Survey Plates of Mount Palomar are searched and, hopefully, the radio source can be tied to an optical object.

If this optical object is small and starlike, observers using either the 200-inch Mt. Palomar telescope, the 120-inch at Lick, or the 84-inch at Kitt Peak, will try to photograph its spectrum and measure the red-shift.

Quasi-stellar sources are a rather remarkable example of the recent results of radio astronomy. Much of the work in radio astronomy perhaps lacks the same dramatic impact, but it is equally important. There are many galaxies within our universe, of a wide variety of forms. Galaxies vary considerably in their brightness at radio wavelengths. Some are abnormally bright while others, like our near neighbor the Andromeda Nebula, appear to be about equivalent to our own Galaxy as radio emitters. Although the brightest of the radio galaxies do seem to include many which have various irregular optical features, there is as yet no good explanation of the relationship between the radio and the optical behavior.

For some of our neighboring galaxies at not too great a distance, radio telescopes such as the 300-foot telescope at Green Bank are large enough and precise enough to be used to detect the radiation from the neutral hydrogen gas within the galaxy itself. By measuring the radiation from this hydrogen we can find the amount of the gas, the way it is rotating in the galaxy, and the apparent velocity of motion of the galaxy as it moves with respect to the Earth.

Our own Galaxy itself has been mapped with great care to find how the neutral hydrogen gas is distributed within it. The results of these radio measurements have shown in a very clear way the spiral structure of the Galaxy. The center of our Galaxy cannot be seen optically, but radio observations have located many complex regions of radio emission in the nucleus and given the first evidence of motions of matter in this part of the system. Here again one sees the advantage of the longer radio wavelengths rather than light waves because the medium within the Galaxy absorbs and scatters radio waves less than is the case with ordinary light.

For several years radio astronomers were able to observe only one spectral line: the famous 21-centimeter wavelength radiated and absorbed by neutral hydrogen gas in interstellar space. Various other lines have been predicted and searches made for them. Recently, the complex set of lines associated with the OH radical—the combination of an oxygen and a hydrogen atom—has been observed. Like so many discoveries, it brings new and unexplained problems in its wake. The intensities of the lines in the series are not what were expected. The radio waves are polarized and there are regions in the sky where the intensity of the emission varies with time. These phenomena are not yet explained.

Even more recently workers in the Soviet Union and the United States have identified another series of lines from the hydrogen atom. These arise from excited hydrogen, and thus are seen best in the hot masses of gas near bright stars. Their existence allows us to measure the velocity of the excited hydrogen gas and the electron density within it.

It would be possible to continue to list other areas of astronomy where the radio observations have brought new results: in the study of bursts of radio emission from the Sun and from the planet Jupiter, for example, or the estimates of the Venus surface temperature made from its radio brightness. It is probably better to inquire what the future may hold for research in radio astronomy.

It is clear that the science is still in the growing state. Many observations which need to be made are practically possible with existing instruments. Simple measurements of the positions, sizes, and intensities of radio sources are still far from adequate. Many of the sources appear to have structure which gets more complex as resolution improves. The variability of some sources is of great interest; in the simplest case of the supernova remnant known as Cassiopeia A, we have a gradual slow decline in the source intensity. But some of the quasi-stellar sources may change intensity by measurable amounts within weeks. Such rapid changes are usually associated with rather small objects; it is generally agreed that variability sets an upper limit on the physical size. However, if the measured red shifts of the quasi-stellar sources are interpreted as a distance measure, their small size derived from the variability leads to problems of how the energy is generated.

The making of more observations is an essential to any larger plan for the future of radio astronomy. One of the possible tasks is the study of the radio sources at great depths in the universe. The small scattering and absorption of radio waves may permit observations of very distant radio sources. The statistics so gathered can help answer the questions of the early history of the universe. Already it seems, as we examine the numbers of radio sources as a function of their intensity, that there is an excess of the weak radio sources. A start has been made in collecting statistics on the angular sizes of radio sources and also on their radio spectra. These observational facts for the most distant sources must, of course, be consistent with cosmological theory. As yet the observations are not complete enough to make definite choices between one cosmological theory and another, but such a possibility may not be far distant.

There are always astronomical problems associated with evolution, for we essentially have been able to observe so far only for a tiny flash of astronomical time. If the universe is evolving and changing, it may be that our brief picture shows us, when we look to great depths in space, various steps in its evolution.

Lastly, there is the ever present chance of discovery. We have so much to do in radio astronomy that has not yet been done that the unexpected result must always be a possibility. Experience already shows that it is not necessary to plan to discover something new in radio astronomy in order to find it.

To continue and extend radio astronomy requires new telescopes and techniques as well as new ideas. The rapid growth in the design and improvement of our ground-based telescopes shows no signs of slowing down. We also see the beginnings of space radio astronomy. Some observations have already been made from spacecraft, and during the next few years we shall see a considerable extension of space radio astronomy.

• 24

The X-Ray Universe

HERBERT FRIEDMAN

Of all the sciences, astronomy stands to gain the most from man's newly acquired ability to send his scientific instruments into space on rockets and satellites. Above the atmospheric blanket that envelops the Earth, the full range of celestial radiations from ultra-short X-rays to ultra-long radio waves is exposed without attenuation, and no basic limit is imposed on the astronomical capability to resolve the smallest heavenly bodies. By contrast, the earth-bound astronomer can penetrate the murky atmosphere only through two narrow windows—the range of visible light between 3,900 and 7,600 angstroms and the band of radio waves from about 1 centimeter to 40 meters. And through the shimmering atmosphere, all celestial objects twinkle and blur.

In the first two decades of rocket astronomy, ultraviolet and X-ray observations have focused principally on the Sun. Much has been learned of the energetic processes that work their way through the solar atmosphere and flood the planetary environment with plasma clouds, ionizing X-rays and ultraviolet light, and particles of cosmic ray energies. Although major discoveries have been made in solar astronomy, none match the complete surprise that has come in recent years with the detection of mysterious

X-ray sources far beyond the solar system. From 1949, when X-rays were first observed from the Sun, thirteen years elapsed before the potential of cosmic X-ray astronomy became apparent. Within the past few years about twenty-five discrete sources have been discovered, and we can estimate from this sample that *X-ray stars* may be as abundant as *radio stars* numbering in the tens of thousands.

In many ways, the progress of discovery in X-ray astronomy parallels the early history of radio astronomy, and offers every prospect of providing equally profound revelations of the nature of the universe. Some of the cosmic X-ray sources may form a new class of astronomical objects, powerful X-ray emitters that are undetectable in visible or optical wavelengths. We still have no clear understanding of the nature of any of the X-ray sources thus far discovered.

Interstellar space is not entirely void. It is permeated by a very thin gas, about 1 atom per cubic centimeter compared to 10^{19} atoms per cubic centimeter in the air that surrounds us, and a sparse sprinkling of dust, just a few grains per cubic kilometer. Even this very dilute medium becomes opaque over the distances to the nearest stars at ultraviolet wavelengths which fall just short of the limiting wavelength that will ionize the hydrogen atom, 912 angstroms. As the wavelength decreases, interstellar space becomes gradually more transparent. But not until the wavelength is shorter than 10 angstroms in the X-ray region can the rays traverse the distance from the center to the edge of our disk-shaped Galaxy. X-ray astronomy is, therefore, primarily concerned with the wavelength range smaller than 10 angstroms. In energy terms, the range is approximately 1,000 to 100,000 electron volts.

To observe the softer X-rays, it is necessary to exceed 100 kilometers—a height which is readily accessible to small research rockets such as the Aerobee-Hi. Harder X-rays are more penetrating and, at balloon altitudes—about 30 kilometers—cosmic X-ray wavelengths shorter than 0.5 angstrom are observable.

Most of the observational data have thus far been obtained with detectors flown in Aerobee rockets whose flight time above 100 kilometers does not exceed 5 minutes. Some of the rockets have been stabilized so as to aim their detectors in selected directions, but most of the observations have been made with uncontrolled rockets that spin and precess freely in space. The solution of the aspect problem—that of determining how the rocket is oriented at any instant of time—is derived from signals produced by optical star sensors, horizon sensors, and magnetometers.

X-rays cannot be refracted appreciably, so that image-forming lenses are impractical. Neither can X-rays be reflected in the conventional way by mirror telescopes. Reflection is possible only at nearly grazing angles with a mirror surface. In principle, an X-ray telescope can be built in the form of a paraboloidal surface which approximates closely to the interior of a cone of very small vertex angle. Such telescopes, about 10 centimeters in diameter, have already been used successfully to photograph X-ray images of the Sun. However, it requires a very large aperture when viewed at near-grazing angles and a very long focal length to provide a large X-ray "gathering" power. Reflecting X-ray telescopes will eventually be flown in large orbiting observatories such as can be launched with Saturn-class vehicles. But until now, cosmic X-ray telescopes have consisted of nothing more than mechanical baffles to limit the field of view of conventional X-ray detectors such as Geiger counters, proportional counters, and scintillation counters.

The search for cosmic X-rays began in 1956, when James Kupperian and I flew a scintillation counter on a small rocket and observed a diffuse emission from above the atmosphere in the range 20 thousand to 100 thousand electron volts. We followed with several unsuccessful attempts to detect localized sources of emission in the next few years. The first convincing evidence of the existence of discrete sources of cosmic X-rays beyond the solar system came in 1962, when R. Giacconi, H. Gursky, F. R. Paolini, and B. B. Rossi observed, with a broad field of view, what appeared to be a large region of X-ray emission in the general direction of the galactic center. This exciting discovery stimulated more ambitious efforts and, a year later, my colleagues and I at the Naval Research Laboratory were able to localize a very strong source in Scorpius and a second source about one-eighth as bright in the direction of the Crab Nebula in the constellation Taurus. The source which we designated "Sco XR-1" was 22 degrees above the galactic plane and must have been responsible for the major portion of the poorly resolved signal detected in 1962.

It was, indeed, amazing to find a discrete source as bright as the X-ray Sun in the same range of wavelengths. Even if it were as close as 100 light-years, its intrinsic X-ray brightness would need to be 10 million million times that of the Sun. To add to the surprise and the mystery, there is no clearly related optical or radio emitting source in the direction of Sco XR-1. In contrast, the Crab Nebula is one of the most spectacular optical and radio phenomena in the heavens. It is the remnant of a supernova, a star that exploded in A.D. 1054, and the debris of the explosion has since then been expanding in space at about 100 kilometers a second, so that it now stretches across 6 light-years.

In visible light, the Crab Nebula is composed of an amorphous core of white light which appears to be enmeshed in tangled filaments of a gas that glows red in the characteristic light of excited hydrogen atoms. The light of the core, moreover, is observed to be highly polarized, as one would expect from synchrotron radiation—light generated by electrons or protons as they spiral along a magnetic field at nearly the speed of light. Such bluish-white synchrotron light, highly polarized and continuous in its spectral distribution, can be observed in the laboratory to surround the beam of a high-energy synchrotron accelerator. In the radio spectrum, the Crab is the third brightest object in the sky. Its spectrum is continuous and the radio waves are strongly polarized. The radio spectrum and the optical spectrum fit together in a smooth continuous curve, and both can be explained by a common source of electrons with energies ranging from ten million to a million million electron volts, gyrating in a magnetic field of about one ten-thousandth of a gauss. If we extrapolate the optical spectrum toward the short-wavelength X-rays, the predicted flux is within an order of magnitude of the observed. To produce the X-ray emission, it would be necessary for the electron energy range to reach as high as 100 million million electron volts or, alternatively, that there be knots of much more intense magnetic field into which lower energy electrons could drift and be forced to gyrate in tighter spirals.

With two such remarkably different sources as Sco XR-1 and the Crab Nebula to explain, a variety of interesting theoretical speculations were quickly offered. The X-ray luminosities involved were of such a magnitude that only the most energetic processes known in astrophysics could be responsible. Within the galaxy, supernovae are certainly the most violent phenomena observed. Could the source in Scorpius, like the Crab X-ray source, also be related to a supernova? If so, why is there no bright radio or optical nebulosity to match the X-ray source? One possibility is that Sco XR-1 is a neutron star.

It is generally believed that a supernova evolves from a common star somewhat more massive than the Sun, which lives through successive stages of nuclear burning until it has synthesized a core made up largely of iron-group elements. When the iron core reaches a temperature in excess of a few billion degrees, it disintegrates into neutrons and alpha particles and collapses catastrophically to a density comparable with that encountered in the atomic nucleus. The collapse ends in a compacted star only 10 kilometers in radius at a density of about 1,000 million million grams per cubic centimeter.

This hypothetical neutron star is a purely theoretical concept and no evidence for its existence has ever been obtained. We would expect it to have a photosphere only a few meters thick, which would separate the 1,000 million-degree core from interstellar space, and in this thin skin the temperature would drop rapidly to about 10 million degrees Kelvin in the last centimeter of surface. The gaseous atmosphere at the fringe of the star would consist of relatively normal atoms radiating a continuous black body spectrum with its maximum concentrated in the 1–10 angstrom region, and peaking near 3 angstroms. The tail of emission reaching into the visible spectrum would contain so little energy that it would be wellnigh unobservable even with a large optical telescope. The neutron star model, therefore, seemed a promising explanation for Sco XR-1, the bright X-ray source with no accompanying visible or radio emission.

It is further believed that, before a star becomes a supernova, its core is surrounded by an envelope containing appreciable amounts of unburned, lighter nuclei. When the core collapses, this envelope also falls in and incidentally gets heated to a temperature at which it ignites in a thermonuclear fashion, and there is a violent explosion. The debris spreads rapidly into space to form a nebula such as we now see in the Crab. The neutron star proponents suggested that the X-ray source in the Crab could also be a neutron star rather than the surrounding nebula.

A third possible explanation of the cosmic X-ray sources is that they are produced by bremsstrahlung in dilute, hot gas clouds at temperatures from 10 to 100 million degrees Kelvin. It is the braking of fast electrons by collisions with atoms that generates the bremsstrahlung X-rays. Such hot clouds would concentrate their radiation in the X-ray region and emit a relatively insignificant amount of visible light and radio waves. The solar corona is a well-known example of such a hot plasma with temperatures that normally range up to a few million degrees. It produces a continuous bremsstrahlung spectrum on which are superimposed various spectral lines of the heavier elements. At times of solar flares, the active region of the corona may be heated to hundreds of millions of degrees, and X-rays are emitted up to energies of several hundred thousand electron volts.

To narrow down the choice of models, it is crucial to determine whether the X-ray sources are small—of stellar dimensions, less than a second of arc in diameter—or whether they are extended clouds such as the visible Crab. If a source is extended, the neutron star explanation cannot apply. With the X-ray baffles now used to define the field of view, it is not possible to determine the size to better than about 0.1 degree of arc.

A similar problem has faced radio astronomers whose telescopes have large flux-gathering power but relatively poor angular resolution. They have been very successful, however, in obtaining accurate position and size data from observations of occultations of radio sources by the Moon. As the Moon travels around the Earth, it eclipses the stars within a narrow belt of the celestial sphere. Radio astronomers set their telescopes on a source and observe the time and rate of disappearance or reappearance as the Moon passes between. A star vanishes abruptly, an extended source relatively slowly. This principle has been applied to determine the size of the X-ray source in the Crab Nebula.

To observe an X-ray source occultation from a rocket is considerably more difficult than to observe a radio occultation from the ground. The duration of flight is a matter of minutes, and the Moon traverses the sky at the rate of half a minute of arc per minute of time. It would take 12 minutes to eclipse the full Crab Nebula, yet the rocket could remain aloft only 5 minutes.

We therefore decided to launch the rocket at a time which would permit us to observe the occultation of the central region of the Nebula over about 2 or 3 minutes of arc. The observation was planned for July 7, 1964. Another opportunity would not occur until 1972. Fortunately, the observation was carried out successfully. It showed that the X-rays came from a region about one light-year in diameter—roughly one third of the size of the visible Nebula—and centered on it. Clearly, the Crab X-rays are not generated in a neutron star.

If the Crab source is not a neutron star, how well can its X-rays be explained as synchrotron radiation? There is a major difficulty with the synchrotron explanation. As the electrons radiate, they slow down. The slowing down may take several thousand years for the 10 million electron volt electrons that produce the radio waves. Such electrons could, therefore, persist from the original explosion 912 years ago. However, the X-ray-producing electrons slow down very much faster, so fast that they could hardly survive more than a few tens of years, or even less than a year if the magnetic field is as high as a thousandth of a gauss in the central region. Balloon measurements of X-rays of energy 20 to 100 thousand electron volts from the Crab place an even more stringent restriction on the lifetimes.

For the third explanation, the hot plasma, we must find a source of energy that can maintain the nebular temperature as high as 100 million degrees Kelvin to this late date after the explosion. Radioactive decay has been proposed; estimates of the total radioactive energy produced in the original explosion are consistent with the energy output in X-rays.

Within the past two years the Naval Research Laboratory surveys have expanded the list of X-ray sources observed to more than twenty, which lie in the direction of the plane of the Milky Way, and three high-latitude sources which are probably extra-Galactic. Only four of the sources lie in the directions of well-known objects: these are the two distant radio galaxies, Cygnus A at 700 million light-years and M-87, or Virgo A, at about 35 million light-years, and two Galactic supernova remnants, the Crab Nebula and Cassiopeia A. Since we know the distance to the Crab Nebula, about 3,500 light-years, and to Cassiopeia A, about 10,000 light-years, we can calculate their luminosities from the observed X-ray fluxes near the Earth. Both turn out to be nearly equal, about 4×10^{26} kilowatts. The flux from the more distant Cassiopeia A is near the limit of X-ray detectability at the present time with rocket-borne instruments. We, therefore, surmise that sources more distant than Cassiopeia A, that is more distant than 10,000 light-years, would be too faint to be detected. The radius of the disk of our Galaxy is about 50,000 light-years. Nearly all of the sources thus far observed appear then to be within a disk only one fifth of the galactic diameter, or one twenty-fifth of the total volume of the disk. We are thus led to the conclusion that the total number of sources in the galaxy is about 20 times 25, or 500.

The frequency of stellar explosions in external galaxies appears to be as high as one every fifty years. In our own Galaxy, only three have been observed in the last thousand years. But obscuration by interstellar dust may hide the majority of such events. If X-ray sources are related to supernova events, if supernovae occur only once every fifty years, and if there are as many as 500 X-ray sources in the Galaxy, then the average life of an X-ray source must be about 25,000 years. This is much greater than the lifetime for radio emission from a supernova and we can, therefore, understand why we detect so many more X-ray sources than radio remnants of supernovae.

There is evidence of variability in the X-ray sources in the Milky Way. One strong source in Cygnus has been observed to decrease in brightness by a factor of four over a time span of only one year. A complex of X-ray sources along the galactic equator and concentrated towards the galactic center has been observed at various times over the past two years by several groups of experimenters. The positions and intensities mapped by these different groups at different times do not agree very well. This disagreement may in truth reflect a variation in the brightness of the sources. If variability is a common feature of galactic sources, it may provide a clue to the nature of these sources. For example, if we consider the neutron star, it theoretically cools rapidly until it reaches a temperature of only a few million degrees. The time to cool from 10 million degrees to the million-degree range may be of the order of a year. However, it would be a relatively rare change if a neutron star were observed in its rapidly cooling phase.

Another possibility is that some of the X-ray sources may be ordinary novae rather than supernovae. About forty novae per year flash forth in the Galaxy. Within the distance limits of observable X-ray sources, we could expect one or two novae per year. It is possible that the X-ray flux from a nova during its first year would be in the range of the X-ray emissions observed thus far, and the lifetime of this emission would be of the order of a year.

The observations of X-ray emission from radio galaxies present us with theoretical problems entirely different from those of the

Milky Way sources. The two radio galaxies, Cygnus A and M-87, are among the most prolific radio generators in the universe. Great difficulty is encountered in explaining the radio luminosities in terms of synchrotron emission because the energies involved are comparable to the entire thermonuclear energy content of a galaxy. Yet the X-rays we observe from the directions of these two sources are one to two orders of magnitude more intense than the radio fluxes. If it were assumed that the X-ray emission also is part of the synchrotron process, the theoretical difficulty in explaining the energy source would be enhanced by another factor of ten or one hundred.

We must attempt to find other sources of X-ray emission than synchrotron if we are to escape this dilemma. Thomas Gold of Cornell University has suggested that quasars and also radio galaxies derive their energy from stellar collisions in very dense galactic nuclei. The relative velocities involved in these stellar collisions are such as to generate X-rays in the range we observe. According to such a model, the X-ray emission would dominate all other electromagnetic radiations.

The range of intensity thus far observed in X-ray astronomy is about a factor of 100. In the radio spectrum, about 10,000 radio galaxies have been identified and the intensities cover a range of about 10,000. By analogy we would expect that if we could increase the sensitivity of X-ray detection by a factor of 100, we would probably detect 10,000 X-ray sources. All that is needed to achieve the X-ray sensitivity is the longer observing time attainable from a satellite. One hour of pointed measurement at a source would provide a hundred times the sensitivity that is obtained in observations from an unstabilized rocket. On Apollo class vehicles it is not impracticable to consider counters of 100 square meters area capable of mapping the entire sky at one degree per second with a sensitivity adequate to detect sources a thousand times weaker than the Crab Nebula. The sensitivity could be increased by slowing the scan rate. From a base on the Moon, star rise and star set observed over the lunar horizon could provide position accuracies of one second of arc on sources hundreds of times weaker than the Crab. X-ray reflecting telescopes have already been employed successfully to photograph the Sun. In principle, it should be possible to engineer X-ray telescopes as large as can be accommodated in the Apollo Moon vehicles and still achieve a resolution in the range of seconds of arc. Such telescopes would reach X-ray sources a million times weaker than the Crab Nebula.

Within the next decade, X-ray astronomy should become a powerful tool for exploring those regions of the universe where charge particles of very high energy are being generated and where superhot stars and plasmas may exist. To appreciate fully the possibilities of X-ray astronomy, we need only to think of the comparison with radio astronomy. In a quarter of a century of radio exploration man has advanced his knowledge of the observable universe incalculably. Suppose, for example, that radio waves could not penetrate the atmosphere and that our radio astronomy had to be conducted from rockets and satellites. Knowing what we now do of the discoveries of radio astronomy, we would exert every possible effort to conduct the observations above the atmosphere. Today, X-ray astronomy appears to have every ingredient of potential scientific revelation that has characterized radio astronomy, and we are prepared to explore it with the full capabilities of space technology.

• 25

The Cosmic Ray Universe

PETER MEYER

All our knowledge of the physical processes which occur in our Galaxy or even further away in the universe comes from observations of the radiation which is emitted from stars and other astronomical objects and reaches the vicinity of the Earth with sufficient intensity to be observable. Electromagnetic radiation in the visible spectral region—light—is most easily accessible to observation and has for a long time been the only means of gaining information about astrophysical phenomena. Astronomy is one of the oldest branches of the sciences, mainly because some astronomical phenomena are directly accessible to observation without the use of complicated instrumentation. The invention of the telescope has made astronomy one of the most advanced scientific disciplines. To this day astronomical observations are the richest source for new discoveries in astrophysics, and it is amazing what wealth of information about many details in our universe has been gained through the analysis of optical and spectroscopical data.

Visible light, however, is not the only form of radiation which reaches us from the great distances. Twenty years ago it was shown that radio waves of extraterrestrial origin are reaching the Earth, some coming from distinct sources, others from larger regions of space. This discovery led to the new branch of astronomy, radio astronomy. Then, only a few years ago, came the startling discovery of X-ray emitters within our Galaxy. But we should not forget the corpuscular radiation which was discovered more than fifty years ago and which was given the name of cosmic radiation. It is this radiation which is the topic of this chapter.

Although discovered quite some time ago, cosmic rays have only recently, for a little more than ten years, contributed heavily to astrophysical research. The reason for this delay is that it took a long time to ascertain the nature of this radiation. To clarify this, let me briefly go back to some of the historical background which led to the discovery of the cosmic rays.

At the beginning of the twentieth century many research workers were actively engaged in studying the radiation which is emitted from radioactive substances. The scientists constructed detectors which were sensitive to the radiation from the radioactive substances. They soon noticed that even when all radioactive material was removed from the detector, there remained an unexplained small amount of radiation. It was soon argued that this radiation must come from a general radioactive contamination of the surface of the Earth, and in order to prove this point an Austrian scientist, V. F. Hess, placed his detection instruments in the gondola of a balloon and took a ride with them to quite high altitudes above the Earth. If the hypothesis that the radiation originated at the surface of the Earth were correct, Hess expected that the radiation effect in his instruments should go down the farther he removed them from the surface of the Earth. To his surprise, exactly the opposite occurred. The intensity of the radiation increased with increasing height, and Hess soon concluded that his experiment could only be understood if one assumed that a very energetic type of radiation falls on the surface of the Earth from the outside. This radiation, whose nature was not at all understood at the time, except that it had properties similar to the radiation from radioactive substances, was given the name cosmic radiation.

It took many years until the nature of the radiation was clearly identified. The reason for the difficulty is simply that our atmosphere, which is so transparent for visible light, is not at all transparent for cosmic rays. We know today that the cosmic rays are *nuclear particles,* of great energy, and these nuclear particles have a high probability of collision with the nuclei of oxygen and nitrogen in the upper layers of the atmosphere. The radiation, therefore, that we see at sea level is entirely different in character from the primary radiation: it consists essentially only of the nuclear debris which is the result of the collisions.

The long years of study of these interactions between the cosmic ray particles in the atmosphere were not at all fruitless. On the contrary, they led to a large number of very important discoveries and many of the new unstable elementary particles were discovered during these studies.

In order to learn what information the cosmic rays may bring us from the distant worlds, it obviously is necessary to study them directly, quite aside from their atmospheric secondaries. Technological developments of the last years have helped greatly in this direction. First of all, high-altitude balloons were developed capable of carrying instruments in and above the stratosphere. Then the technology of rockets and of satellites came along and made it possible to investigate many additional details of this radiation. Let me in a few words give some of these details.

The majority of the cosmic rays are protons—the nuclei of the simplest atom, the hydrogen atom. But there also exist heavier nuclei in a periodic system: e.g., helium, lithium, beryllium, boron, carbon, nitrogen and oxygen; and nuclei as heavy as iron have been found. There are also some electrons present in the primary cosmic radiation, and all of the particles have energies which vary over a wide range. Many of them have an energy between 100 and 1,000 million electron volts. This falls within the energy range that cyclotrons and high-energy machines in our laboratories can produce. But some of them have extremely high energy, a billion times the energy of the largest accelerator that has been built in a laboratory today.

Two groups of cosmic ray problems are of particular interest in astrophysics. On the one hand is the question of the origin of cosmic radiation: that is, the question of understanding the processes by which these nuclear particles may gain their tremendous energy and the question of the nature of the objects which are capable of producing such particle acceleration. On the other hand, there is the problem of the interactions of the cosmic rays during their travel: that is, the interaction of cosmic rays with the tenuous matter in the Galaxy or with magnetic fields in the Galaxy or in the solar system. In this second context, one may use cosmic ray particles as probes in order to explore phenomena which otherwise are inaccessible to observation.

Let me first address myself to the problem of the origin of cosmic rays. Anybody who has ever seen a high-energy physics laboratory and an accelerator must be convinced that it is extremely difficult to accelerate nuclear particles to high energy. Nevertheless, we seem to find that nature does exactly that at all times, with high efficiency, and at many places. We know that in the Galaxy there exists an abundant flux of cosmic rays which must be replenished. But we also find the phenomenon of particle acceleration closer to the Earth. Take, for example, the Van Allen radiation belt around the Earth, which is filled with high-energy particles. Or let us look at the Sun which, as we know, does at rare occasions emit high-energy particles when a large chromospheric eruption takes place on its surface. In spite of all these observations, it has not yet been possible to pinpoint clearly the mechanism through which particle acceleration in astronomical dimensions proceeds.

We have good reasons to assume that most of the cosmic rays that we observe at the Earth are of Galactic origin. Magnetic fields in the Galaxy force the charged nuclear particles to move along spiral orbits and they prevent them from leaving the Galaxy along a direct path. These magnetic fields are probably the reason that the particles fall uniformly on the Earth, so their direction of incidence does not give us any information as to the point of their origin. In this respect, cosmic rays obviously behave entirely differently from visible light, which travels along straight paths. If our eyes were sensitive to cosmic radiation rather than light, we would not be able to see a single star or object; the entire sky would look uniformly gray.

Enrico Fermi was the first to develop a theory in which motions of Galactic magnetic fields were made responsible for the acceleration of the cosmic rays. But there are other hypotheses which appear to us even more probable today and which claim that cosmic rays originate at specific points in the Galaxy—namely, in the remnants of so-called supernovae. The phenomenon of a supernova occurs in aging stars which have used up their nuclear fuel, hydrogen and helium, and which are then suddenly exploding, creating a large amount of energy in a relatively short time.

Let us ask the following question: Which objects can with certainty be identified as sources of energetic particles? We find that there are only two kinds. The first is the star nearest to us, the Sun, which, as I have just mentioned, emits high-energy particles during chromospheric eruptions called solar flares. The energy of the particles which the Sun normally emits during such flares is lower than the energy of the average cosmic rays, and it is quite clear that the Sun cannot be a major source of the total cosmic radiation that we observe near the Earth.

The Sun, however, is very interesting to us as a prototype accelerator. First, it might be possible that other stars might show phenomena similar to those of a solar flare and are emitting particles more abundantly than the Sun. Second, the Sun is so close to us that we can hope that a detailed study of the solar flare phenomenon may give us a general clue to the acceleration mechanism which will help us to understand the processes in far removed objects. Third, we can correlate the particle emission of the Sun with the emissions of light, of X-rays, and of radio waves and use these correlations as indicators for a theory of particle acceleration.

The study of solar-emitted high-energy particles has also been very fruitful in other respects. For example, it was possible to measure the distribution of the chemical elements in the solar-emitted particles and thereby to obtain a knowledge of the abundance of the elements in the Sun, completely independent of and different from the data which are available through spectroscopic investigations. Finally, the study of the time dependence of the intensity of solar particles has been of great interest for understanding the configuration of magnetic fields in the solar system. The solar-produced particles are unable to escape from the solar system in a short time. They are stored within the interplanetary space, and the only mechanism that one can propose for this storage are interplanetary magnetic fields. Through a study of the temporal behavior of solar-created high-energy particles, it has been possible to make deductions on the configuration of the magnetic fields in interplanetary space.

The only other object in our Galaxy aside from the Sun which we can clearly identify as a source of energetic particles are the remnants of supernovae. Six supernovae have been observed and recorded in our Galaxy in historic times, and on the basis of these data it appears probable that one or several such supernovae explosions occur per century within our Galaxy. The remnants of the supernovae often show an intense emission of electromagnetic radiation with a spectral distribution which can uniquely be diagnosed as being due to high energy electrons. There is no doubt today that several of the supernova remnants contain high-energy electrons. The most famous of the supernova remnants is the Crab Nebula.

The supernova explosion which led to the Crab Nebula was observed in the year 1054 by Chinese astronomers and was described by them in great detail. While the observations of the visible light, its polarization, and the radio waves which are emitted by the Crab Nebula leave no doubt about an intense flux of high-energy electrons, there is unfortunately no direct proof for the existence of energetic protons and heavy ions, but it has been assumed likely that those particles have also been accelerated. It is, of course, not known to what degree the particles which are present in the supernova remnant are emitted into the Galaxy and are therefore becoming part of the general cosmic ray gas in the Galaxy.

While I said that we think that most of the cosmic rays are of galactic origin, we also know today that not *all* the cosmic radiation can be stored within the Galaxy. In recent years, through investigations of large air showers, it has been possible to show that there exist primary particles with energies up to 10^{20} (100 billion billion) electron volts. The flux of these energetic particles is very small, but their very existence is of great astrophysical importance. It forces us to the conclusion that cosmic radiation, probably with a small density, is also present in extragalactic space because it is inconceivable that the magnetic fields of our Galaxy are capable of confining such energetic particles within the dimensions of the Galaxy. We know nothing about the origin of these extremely high energy particles, nor do we know anything about their spatial distribution in extragalactic space. All discussions of this problem are rather speculative at the present time.

Let me then turn to the second part of this discussion and try to point out what we may learn through a study of the interactions of the cosmic rays during their life and travel. As I mentioned before, the trajectories of the charged cosmic ray particles are determined by the galactic magnetic fields. Except for the particles of extremely high energy, the radii of curvature of the particle orbits are small compared to the dimensions of the Galaxy. Therefore cosmic ray particles move through the Galaxy in a random-walk fashion because they are bent around by the magnetic fields, which change in direction and intensity from place to place. Although individual sources may be responsible for the creation of the cosmic rays, the direction of motion of the particles very soon becomes isotropic. They remain in the Galaxy for intervals of the order of a million or 10 million years until they escape into extragalactic space.

We do not know the lifetime of the cosmic rays with great accuracy, but we can estimate it by using our knowledge of the average density of gas in the Galaxy. This density is estimated to be approximately one or two hydrogen atoms per cubic centimeter. The second quantity which we need in order to estimate the lifetime is the average amount of matter that a cosmic ray particle traverses until it is lost through escape.

This second quantity can be obtained experimentally through the observation of the mass spectrum of the cosmic radiation. To point out how this can be achieved, let us recall the following: The light elements, lithium, beryllium, and boron, are extremely rare in the universe and so we can assume that freshly accelerated cosmic radiation also contains very few of these elements. Yet a small but measurable quantity of these elements is observed in the beam of the cosmic rays when it reaches the Earth, a quantity which, relative to the other elements, is very much larger than in the universe. Their existence can, however, be explained by assuming that they are the products of collisions of cosmic ray nuclei heavier than the lithium, beryllium, and boron with the hydrogen within the Galaxy. During such collisions, spallation products are created, and among these the lithium, beryllium, and boron atoms are not rare. By measuring the flux of the light elements and comparing it to the flux of heavier elements, one can therefore estimate how many collisions have taken place and thus conclude that, on the average, three to four grams per square centimeter of matter have been traversed by a cosmic ray particle until it reaches the observer.

The abundances of the various elements in the cosmic radiation are being studied intensively at the present time. These investigations have shown interesting differences between the element abundance in the universe and the abundance in the cosmic radiation. For reasons that I have just outlined, the cosmic rays are relatively rich in lithium, beryllium, and boron.

But if one looks at the heavier elements in the cosmic radiation, one finds that they also show a higher relative abundance than the general abundance in the universe—if one compares them, for example, with the abundance of oxygen. This observation has sometimes been interpreted as a further suggestion that the supernovae explosions are the source of the cosmic rays. I mentioned before that the stars which undergo the supernova process are old stars which have used up a large fraction of their nuclear fuel and in which the heavier elements show the highest concentration.

An important further step in the study of the composition of the cosmic rays is an investigation of the distribution of various isotopes. The measuring instruments that we have at our disposal today are not yet capable of resolving most of the different isotopes. The only successful work has been carried out on the isotopes of the element helium—the isotopes He^3 and He^4—and the results of this work further confirm that the cosmic rays have, on the average, traversed 3 to 4 grams per square centimeter of matter between their source and the Earth.

The method of using the cosmic rays as probes to investigate astrophysical problems has been particularly useful in the studies of the physics of the solar system. Collisions between cosmic rays and interplanetary gas do not play any important role here because the time which the particles spend in this limited region of space is very short. The interaction of the cosmic ray particles with interplanetary magnetic fields, however, is very noticeable for the observer at the Earth. These interplanetary fields are controlled by the stream of highly ionized gas which is emitted by the Sun. The Sun is the origin of the fields, and the solar wind manipulates them. The interplanetary magnetic fields modify the flux as well as the energy spectrum of the galactic cosmic rays. This modulation is most pronounced for cosmic ray particles of relatively low energy, particles with energy perhaps up to a few million electron volts, but in this energy region the influence of solar activity is very noticeable.

The most famous of these so-called solar modulation effects of the cosmic radiation is the 11-year variation of cosmic ray intensity, which was first observed by Scott Forbush. The very fact that the modulation occurs with a periodicity of 11 years made it clear that solar activity, which undergoes an 11-year cycle, must be responsible for it. The intensity and energy spectrum of the cosmic rays have been thoroughly studied during the past solar cycle, which included the period of the International Geophysical Year, and it has been found that in the years of highest solar activity the cosmic ray intensity is considerably reduced compared to the years of low solar activity. This anti-correlation between solar activity and cosmic ray intensity makes it very clear that one is dealing with a modulation effect and not an effect of solar production of cosmic rays. While a few years ago the main effort of the study of cosmic ray intensity variations was carried out with monitoring stations on the surface of the Earth, we have today almost continuous surveys of the cosmic ray flux and energy spectrum through instruments which are carried aboard satellites and space probes. The observations of the cosmic ray flux and spectrum have led to the conclusion that a solar controlled mechanism is at work which efficiently shields the inner solar system from the full beam of cosmic rays. The earlier cosmic ray observations already have indicated that this shielding is carried out by interplanetary magnetic fields which are controlled by the solar wind, and this conclusion has today been beautifully verified through direct measurements of the plasma flow and of the magnetic fields in interplanetary space.

It was possible in the past few years to gain a rather complete picture of the physical phenomena in the solar system. Many details, of course, are still missing and much work remains to be done to understand these details. If we briefly summarize our discussion, we may say that the new field of particle astronomy, which we may call cosmic ray studies, has given a new dimension to astrophysical investigations. The synthesis of observations from various branches of astrophysics has greatly enlarged our knowledge of the physical phenomena which take place in the space that surrounds us. This is true for the regions of space close to us and those extremely far from us. Let us again name some of the problems, perhaps in the order of increasing distance from the Earth.

The Earth is surrounded with a belt of intense radiation of energetic particles, the Van Allen radiation, whose origin is a problem which has not been uniquely solved today. There is, however, little doubt that the energy given to the particles in the radiation belts stems from the Sun and is transported through the plasma wind to the vicinity of the Earth. The acceleration mechanism probably has to be sought in the interaction between the solar plasma wind and the geomagnetic field. We find particle acceleration on the surface of the Sun in connection with solar flares. Here we observe the acceleration process in our back yard, so to say, and we hope that by studying the flare phenomenon in great detail we may understand the processes which are going on in objects which are farther removed and not so easily observable. Investigations of the modulation of galactic cosmic rays in the solar system have shown that there exist interplanetary magnetic fields which are influenced by the solar plasma stream. It has been possible to make theoretical models for the configuration of interplanetary magnetic fields from the studies of the cosmic ray modulation.

At greater distance still, the cosmic radiation is a rather important part of our Galaxy. Its energy density is comparable to the energy density in the form of turbulent motion and magnetic fields, and the energy flow of cosmic radiation onto the surface of the Earth is about the same as the flow of energy in the form of starlight. We find it likely that most galaxies of a type similar to ours contain cosmic rays, and through radio astronomical observations it has been possible to prove that other galaxies at least contain high energy electrons. We believe today that the majority of the cosmic rays which we observe at the Earth have been created within our Galaxy. Several sources may be responsible for this radiation, but a very probable candidate as a source is the supernova. The discovery of particles with extremely high energy has forced us to the conclusion that a small fraction of the cosmic rays comes from regions of space outside of the Galaxy, but we do not have much information as to their source.

The field of cosmic ray research has in the course of the years developed more and more from a branch of high-energy physics to a branch of astrophysics. Most of the new elementary particles which were discovered in the past decades were first observed in the cosmic radiation, and in this way cosmic ray research has made a large contribution to high-energy physics. In the years ahead it may well make an equally large contribution to astrophysics.

• 26

Gravitation

ROBERT H. DICKE

The planetary motion in our solar system, the vortex motion of the galaxies, and the expansion of the universe are dominated by a single universal force, gravitation, so indiscriminate that it tugs at all matter in substantially the same way. Equal amounts of matter or energy in the form of gold or hydrogen, or even electromagnetism confined to a box, are attracted gravitationally with equal strength. It is an enigmatic force that has long puzzled man—a force whose origin appears to be related to the nature of space itself. It is a force closely related to the inertial pull experienced by an observer in an accelerated laboratory. It is a force far weaker than any other that has ever been observed.

It may seem strange to call gravitation weak, for that enormous chunk of matter, the Earth, requires a pull of 3.6×10^{21} kilograms, that is a weight of close to four thousand million, million, million kilograms, to keep it accelerating toward the Sun in the manner demanded by its orbit. This is clearly an enormous force. However, on an atomic level gravitation is extremely weak. For example, the electrical interaction between the electron and proton in the hydrogen atom is 10^{40} times as great as the gravitation interaction.

This strange number, 10^{40}, one followed by 40 zeros (10 thousand million, million, million, million, million, million), is a number so large as to defy the imagination. It can be visualized by noting that the distance out to the farthest visible limits of the universe is greater than the diameter of the extremely tiny atomic nucleus by this same factor. It might also be noted that the total number of atoms in the visible region of the universe is approximately 10^{40} squared—10^{80}.

Have scientists an explanation for the weakness of gravitation? Most physicists seem to regard the extreme weakness of gravitation as a result not requiring explanation. According to them, the number 10^{40} cannot be predicted from theory, and it is not related to any of the other numbers with which they deal.

A small minority of astronomers and physicists, of which the famous British astronomer Sir Arthur Eddington was the leader, believed that the number 10^{40} could be obtained from theory, being related in some complicated way to such prosaic mathematical numbers as 2, π, and e. Another small group of physicists believes that the strength of the gravitational interaction may not be fixed but may be related to the structure of the universe.

In 1937, the British physicist P. A. M. Dirac suggested that the apparent relation between the strength of the gravitational interaction, the distance to the visible limits of the universe, and the number of atoms in the visible universe was not an accident, but that these numbers were related to each other. It is known that the age of the universe is approximately 10^{40} times as great as the time required for light to pass through the nucleus of an atom. This suggested to Dirac that as the universe aged and this number increased, all these large numbers would increase together. In particular, in comparison with electrostatic forces, gravitation would become weaker with time.

More recently, a somewhat similar result has been obtained in a quite different way, by analyzing theoretically the role of inertial forces in relation to gravitation. Everyone is quite familiar with inertial forces, for instance, the apparent tug experienced in an automobile as it rapidly rounds a corner. Such forces seem to be intimately related to gravitation. It was suggested by the German physicist Ernst Mach in the nineteenth century that the inertial forces experienced in an accelerated laboratory may be considered as gravitational, having their origin in distant parts of the universe. This idea has become known as Mach's principle.

When Einstein constructed General Relativity, his theory of gravitation, he was strongly influenced by these arguments of Mach, and most of the effects to be expected under Mach's principle are to be found built into his theory. However, there are several striking omissions, and physicists are divided in their estimate of Mach's principle, its significance for physics, and its role in the theory of relativity. Some believe that General Relativity supplemented by an auxiliary condition—a proper initial distribution of matter—is in complete accord with this principle. Others believe the contrary.

A number of years ago C. Brans and I constructed a modified form of Einstein's theory which is able to avoid some of these difficulties without the necessity for supplementary conditions. This theory was subsequently found to be very closely related to one constructed earlier by P. Jordan to provide a theoretical basis for Dirac's ideas. We also were led to expect gravitation to grow weaker with time, but less rapidly than Dirac had suggested. We concluded that the gravitational constant might presently be decreasing by 1 to 2 parts in 10^{11} per year, roughly a factor of 10 less rapidly than Dirac postulated.

This change with time is due to the presence of matter in the universe and the influence of this matter upon the value of a field which has magnitude but not direction—in short, a scalar field. The theory assumes the existence of such a scalar field, the source of a long-range attractive force between matter. This scalar attractive force between matter is very similar to gravitation, and 10 percent of the force we call "gravitation" could be due to a scalar field without the appearance of any obvious anomaly.

The idea of a slowly weakening gravity is not in accord with Einstein's theory of gravitation, and it is of interest to inquire whether it is possible to use modern techniques to determine if gravitation is growing weaker with time. Also, if it should happen that gravitation is weakening, would this have important implications? Would the history of the Moon and the planets in the solar system be affected? The internal workings of the Sun? Of immediate and direct importance to man himself is the question of whether the evolution of the Earth and the origin of life on Earth could have been influenced by such a factor.

Consider first the possibility of using planetary motion to obtain information about the strength of gravitation. If gravitation were growing weaker with time, a gravitational clock would gradually run slower with time in relation to a clock which is based on the electrical interactions. The motions of an electron inside an atom are controlled by the electrical interactions, and a clock based on the internal motions in an atom is called an atomic clock. Such atomic clocks are already available and some are timekeepers of extremely high precision, better than one part in 10^{11}. If a highly precise clock based on the gravitational interaction could be obtained, it would be possible to intercompare the two clocks and hence to determine if the gravitational interaction is growing weaker with time.

It is possible that such a gravitational clock could be constructed as an artificial satellite moving in such a way that the effects of gas drag and light pressure on the motion of the satellite would be negligible. It would be hoped that the timekeeping qualities of such a gravitational clock, when compared with an atomic clock, would yield information about the constancy of the gravitational interaction. As things stand now, it would appear that the artificial satellite would need to be in orbit for several years before enough precision could be obtained to give a definitive answer to the questions which are being raised.

We are tantalizingly close to being able to use the old classical astronomical observations of the Moon and Sun to answer this fundamental question. The Earth, in spinning on its axis, is measuring something akin to atomic time, for the diameter of the Earth is determined mainly by the diameter of the atom. It is not an extremely precise clock, however, as the rotation rate is affected by a variety of disturbances. Because of tidal interactions with the Moon and Sun, the spin of the Earth is gradually decreasing. Also, because of atmospheric tides driven by heat from the Sun, there is a tidal torque tending to increase the rotation rate of the Earth. There is also an irregular fluctuation of the Earth's rotation, the causes of which are not well understood.

As a result of the classical work of astronomers and the recent careful work by the American geophysicists Walter H. Munk and Gordon J. F. MacDonald, the strengths of the tidal interactions with the Earth are now fairly well known and the Earth-clock can be corrected for these effects. The irregular fluctuations are still a source of trouble, but by making use of observations extending over a long enough period of time their contributions tend to average out.

Fortunately, it is possible to combine information concerning the occurrence of ancient eclipses with modern telescope observations in such a way as to be able to compare the timekeeping of the Earth, rotating on its axis. Furthermore, this comparison extends over a 2,000-year period.

The observations permit us to evaluate the magnitude of the tidal slowing of the Moon's motion, hence of the tidal slowing of the Earth's rotation. After making a correction for this effect, the Earth's rotation provides a measure of time. With this time scale, the Earth and Moon are observed to move in their orbits ever more slowly. The slowing of this motion is by roughly twice the amount to be expected if gravitation becomes weaker with time by two parts in one hundred thousand million per year (10^{11}).

While this is an interesting result, one should not be overly impressed. The change of sea level in the past few thousand years requires a further correction to the Earth's rotation. While this appears to be a small correction, even its sign is somewhat uncertain. If the deep interior of the Earth flows easily, the correction for the effect of sea level change decreases the planetary slowing slightly. If the deep interior is rigid, the implied planetary slowing is increased. Another point at which there is some uncertainty concerns the liquid core of the Earth. Its rotation could conceivably be decreasing slightly and speeding the rotation of the outer parts of the Earth.

While there is presently no clearly defined alternative explanation for the observed anomaly in the Earth's rotation in relation to planetary motion, the interpretation as an effect of slowly weakening gravitation is somewhat shaky. It seems certain that some new observational tool will be needed to answer this question.

In addition to the timekeeping artificial satellite mentioned above, as an extremely precise gravitational clock, the Moon itself could be used. The precision of the observations could be greatly improved by placing one or more corner reflectors of great precision on the Moon's surface. By observing the time delay in laser-light pulses returned from these reflectors, a very precise lunar orbit could be obtained. Instead of the rotating Earth, a hydrogen maser would be used as an "atomic clock."

Presently, the most important test of Einstein's gravitational theory involves the motion of the planet Mercury. The major axis of this elliptical orbit swings slowly about the Sun, and part of this motion is very likely due to a relativistic effect. There is some doubt about the magnitude of this effect because part of it could be due to the distorted gravitational field associated with a very slightly flattened Sun. As of now as much as 15 percent of the excess rotation could be due to this non-relativistic effect, the remainder being relativistic. At Princeton, M. Goldenberg and I are attempting to measure this flattening or set an upper limit to its size. Observations in the summer of 1966 indicate that 8 percent is probably due to the effect of the Sun's oblateness.

We come now to the question of the possible effects of a time-varying gravitational interaction on the past history of the planets, the Sun, and the Galaxy. The rate of weakening of gravitation contemplated is so small that great periods of time are required to produce any important effects. The magnitude of change considered reasonable is one or two parts in 10^{11} per year. During all of recorded history the change in the gravitational interaction, resulting from such a small rate of change, would have been insignificant, but during the 4.5 thousand million years that the Earth has existed the change would have been as great as 10 percent, a small but not insignificant change.

The Earth is compressed substantially by the gravitational pull that holds it together. As first pointed out by Pascual Jordan, a steady weakening of this pull would result in a slow but steady expansion of the Earth. It seems possible, perhaps even probable, that this small effect would have left no visible trace, because of larger masking effects produced by a slow convection of the Earth's interior. However, it is possible that steadily weakening gravitation could be a factor leading to the tremendous outpourings of lava on the Earth's surface. It is the deep interior of the Earth that is gravitationally compressed, not the surface, and it is the interior that would expand. This expansion could take place as a percolation of the more fluid parts of the Earth's interior through the crust to the surface. It is interesting and may be significant that the total volume of the Earth's crust is two-thirds of what would have been expected if an expansion had taken place solely in this way, with gravitation weakening at a rate of 1 times 10^{-11} per year for 4.5 thousand million years. (An alternative mode of expansion is the opening of surface cracks.)

The Moon shows little evidence for the internal convection that would lead to the formation of ranges of folded mountains and sideway slipping of large masses along fault planes. On the Moon, therefore, although the expected expansion is far smaller than that for the Earth, it might be more readily observed. The expansion could take place through tension cracks at the surface or again as an outpouring of lava on the surface. Interpreting the lunar maria as lava flows arising in this way, one would expect an outpouring sufficient to cover one quarter of the lunar surface to a depth of one kilometer.

One important effect of a gravitational interaction stronger in the past would be a hotter Sun. It is expected that a star having a mass about that of the Sun would radiate heat and light at a rate proportional to the seventh power of the gravitational constant. This higher radiation rate, some one to two billion years ago, could have meant a warmer surface for the Earth and could have implied a slightly warmer environment for the conditions under which life arose. On the other hand, the traditional view, based on the assumption that the gravitational interaction has not changed with time, is that the Sun and the Earth were somewhat cooler at this earlier time, near the freezing point some 3×10^9 years ago.

One of the ways in which a stronger gravitational interaction in the past would have affected the stars would have been in their apparent age. If the stars were brighter in the past, as a result of the gravitational interaction having been stronger, they would have been evolving more rapidly, running through their life span at a rate more rapid than now. Consequently, their apparent age, as determined now from the observed stage of their evolutionary cycle, could be somewhat faulty, having been computed on the basis of an assumed constant gravitation.

A rather interesting and convincing test of the hypothesis that gravitation is getting weaker with time could be obtained if we had a proper measure of the apparent evolutionary age of the Sun. The Sun's age is already known with considerable precision from the radioactive dating of the meteorites, making use of the radioactive decay of uranium. If, at the same time, one had some direct measure of the degree of evolution of the Sun, one could obtain a quite direct and unambiguous statement about the rate of evolution of the Sun in the past.

The present evidence on the evolutionary ages of stars in comparison with the expansion age of the universe suggests that perhaps the old stars are younger than they appear to be, and this is in agreement with the expectation associated with a gradually weakening gravitational interaction. These conclusions are not firm, however, as the present status of the theory of stellar interiors is not sufficiently well developed that one can be completely confident about its accuracy. It does seem certain that if the gravitational interaction were as much stronger in the past as we think possible, the early days of our Galaxy could have been much different from the situation as it is now pictured. Stars would have been brighter. Their radiation would have been farther into the ultraviolet. The early Galaxy could have spent its store of energy like a wastrel, rapidly evolving in a few hundred million years into a relatively elderly, well developed collection of stars.

To summarize, we have seen that there is a reasonable alternative to Einstein's theory of gravitation and that observations of great accuracy are necessary to provide a crucial test. We have also seen, if the gravitational interaction has been growing steadily weaker with time (the effect of a scalarfield in a matter-filled universe), that this has had important influences on the past history of the solar system, of the stars, and of the Galaxy. In particular, geologists and astronomers would be led far astray if they failed to take into account the effects of such a change in the gravitational interaction, assuming that the change has occurred.

Could such an extremely slow variation of such a weak interaction have important practical consequences for us here and now? The answer appears to be that we do not know. The fact of the matter is that gravitational and inertial forces are very primitive and fundamental. Any lack of understanding here could contribute to a confusion in our understanding of other parts of physics. In particular, we know very little about the structure of elementary particles. Gravitation could conceivably hold the key to their explication.

For a half century there has been very little improvement in the observational basis for our understanding of gravitation. New techniques, particularly those of space science, promise to do much to remedy the situation. In particular, it is to be hoped that in the next half decade we shall discover whether or not gravitation is growing slowly weaker with the expansion of the universe.

• 27

Cosmology

P. J. E. PEEBLES

The purpose of cosmology is to establish a general impression of what the universe is like: What is the nature of the universe, how did it get this way, and how does it end? Put in these terms, it is apparent that this is an ancient and honorable subject for debate. The most recent revolution of thought on the subject has its roots in the second and third decades of this century, with the revision of theoretical ideas provided by Einstein's General Relativity theory and the great change in astronomical observations made possible by the construction of big telescopes. These two developments combined in a remarkable way to give us the presently accepted picture of the universe.

When Einstein attempted to apply his theory of General Relativity to obtain a theory of the universe, he accepted as most reasonable and natural a very important assumption: that the universe is uniform on the average. This means that the universe would appear much the same when viewed from any position in space, if we ignore the minor local irregularities due to the lumpy distribution of matter in stars and galaxies. This denies the notion that there is an edge, or boundary to space, and it also denies that the Earth is in a preferred position in space. The pendulum thus had swung hard over from the cosmology of a few centuries earlier.

In the cosmological model which Einstein constructed space is curved, closed in around itself. A two-dimensional analogy to this is the surface of a balloon. An ant crawling on the surface would discover that it has no boundaries, for it is the same everywhere if the balloon is spherical, consistent with the assumption of a uniform universe. Furthermore, if the ant crawled far enough in one direction he would go all around the balloon, and end up back where he started. This is the meaning of the closure of the space: in Einstein's original model, if a traveler moved in one direction far enough he would end up back at his starting place.

Einstein's model is a static one: in the two-dimensional analogue the size of the balloon is constant. Like the assumption that space is homogeneous, this property was dictated by a philosophical consideration, that the universe should be eternal, unchanging. To make this idea consistent with his theory of General Relativity, Einstein had to introduce a new kind of force, the so-called cosmological term, to balance the attractive force of gravity. Other people showed that with the original theory of General Relativity, without the cosmological term, one could also find models of the universe in which space is closed, as in Einstein's model, but the curvature of the space necessarily is changing with time. In the two-dimensional balloon analogy we must imagine that a small boy is blowing up the balloon, or else letting the air out of it. An interesting property of this model is that if the universe initally were expanding, the expansion eventually would slow and stop, and the universe would collapse back all the way to zero volume. One could not contrive to circumnavigate this closed-model universe in the manner possible in Einstein's original model because the trip necessarily would take so long that before it could be completed the universe would have stopped expanding, collapsing back to zero volume. Neglecting the cosmological term, we know that in addition to this closed model for the universe there is another possible kind, the so-called open model, in which space is curved but not closed in on itself. In this open model, if the universe were expanding initially, it would keep on expanding indefinitely.

These theoretical models were derived with little reference to observation, but the great interest in them is due to a remarkable coincidence of the models with astronomical evidence that was accumulating at about the same time.

This evidence grew from the study of the extra-Galactic nebulae or galaxies. These are bounded systems of stars and gas, containing as many as a million million stars each. We are in such a star system, and so are all the other individual stars we can see in the sky. The nearest neighbor of any consequence is the Andromeda Nebula.

It is more than 10 million million kilometers away, and it stretches over several degrees in the sky. Other galaxies can be distinguished at distances so great that they cover only a few seconds of arc in the sky.

Once it was discovered that these galaxies are large and very distant star systems, it was natural to ask how they are distributed about us. The answer appears to be that on the average they are uniformly distributed throughout space. This result was obtained in a beautifully simple way. We know, according to the inverse-square law, that the observed brightness of a galaxy varies inversely as the square of its distance away. Also, if the galaxies were uniformly distributed through space, the number of galaxies within a given distance of us would vary directly as the volume enclosed—which is to say, as the cube of the distance. Putting these two results together, it is seen that if the galaxies were uniformly distributed, the number of galaxies which appear in the sky brighter than a given value would vary inversely as the 3/2 power of the brightness. This is a relation which can be directly checked by counting galaxies to various limiting brightness, and the galaxy counts agree with the relation.

The second important result was the discovery of the general motion of the galaxies. This was established by means of the Doppler effect, according to which the light from a galaxy moving away from us would appear shifted toward the red end of the spectrum, while the light from a galaxy moving toward us would be shifted toward the blue. It was found that the light from very distant galaxies is red-shifted, indicating that the galaxies are moving away from us. Furthermore, the recession speed of a galaxy was found to be directly proportional to its distance away: more distant galaxies are more rapidly moving away from us. This is Hubble's law of the general recession of the galaxies (1929).

We have from these observations the picture of galaxies uniformly distributed about us and uniformly moving away from us. On the face of it, this would seem to lend considerable support to a sort of geocentric cosmology: surely, if the galaxies are moving away from us in all directions, we must be located in a special, central position in the universe. This argument is misleading, as can be seen by considering how the motion of the galaxies would appear to an observer in some other galaxy. Since the observer's galaxy is receding from us, the observer would say that our own

Galaxy is receding from him. Furthermore, let us compare the observations of a third, more distant galaxy in line with us and the observer. According to Hubble's law the more distant galaxy is more rapidly receding from us. Thus the observer in the intermediate galaxy also would see that the third galaxy is moving away from him, but at a lower speed than we see. At the same time the third galaxy is closer to the intermediate galaxy. This argument leads to the result that, according to Hubble's law, an observer on any galaxy also would see that the galaxies are receding from him, just the same as the general recession we observe.

Hubble's law thus has a simple but remarkable consequence: the motion of the galaxies would appear much the same when viewed from any galaxy. Also, we have seen that the galaxies themselves are uniformly distributed. Thus we have the delightful result that the distribution and motion of the galaxies agree with the predicted behavior of the universe in the uniform expanding cosmological models. Apparently the galaxies are bright enough and distant enough that we can use them to map out the large-scale structure of the universe, and this structure coincides with the cosmological models. It is worth emphasizing that the cosmological models were written down before the observational results were obtained. Our admiration for this feat of pure thought is tempered only slightly by the recollection of the number of false starts which preceded it.

When Einstein learned of these results, he concluded that his introduction of the cosmological term was wrong. Without the addition of the cosmological term, his theory of General Relativity requires that the universe be expanding or contracting: apparently it is expanding, and the original General Relativity theory is adequate to describe its known properties.

In the two-dimensional balloon analogy to this cosmology, the galaxies would correspond to dots painted on the surface of the balloon, the density of dots being roughly the same all over the balloon. As the balloon is blown up the dots move apart, in agreement with the general recession of the galaxies. This analogy is misleading in one respect: the dots would be stretched as the balloon expands, but the galaxies themselves are not expanding. In effect, space is opening up between the galaxies.

Using the cosmological model we can trace the expansion of the universe back in time, ultimately to a time some 7 to 10 thousand million years ago, when the model has shrunk back to zero volume. This striking aspect of the theory has earned it the name the "big bang" cosmology.

This development of fundamental ideas was completed by 1930, and it signals the end of the classical period of cosmology. The cosmological model which was developed is a beautiful combination of theory and observation, but it can be argued with some justice that the observational evidence is not adequate to the theory. The cosmology rests on just two observational results: the uniform distribution and the general recession of the galaxies. The big bang is a long extrapolation from this.

In the past two years a third fundamental piece of evidence has been uncovered, one which seems to provide the direct evidence we have been lacking: that the universe did in fact expand from a highly contracted state. This approach was suggested by R. H. Dicke of Princeton University. It is based on two assumptions: first, that the big bang cosmology is valid and, second, that in the early highly contracted phase the matter in the universe was very hot. The hot material would radiate and fill space with electromagnetic radiation, of a character known as blackbody radiation. The intensity and spectrum of this blackbody radiation are determined solely by the temperature of the material. The same sort of effect is seen in the wall of a stove, which glows dull red when it is heated to about 1,000 degrees Centigrade. Perhaps a more direct analogy is the fireball of radiation produced in a nuclear explosion. In fact, the blackbody radiation which would have been produced in the early stages of the big bang has been called the primordial fireball.

To understand the effect on this primordial fireball of the expansion of the universe, it is well to go back to the two-dimensional balloon analogy. We can picture the radiation as a swarm of ants crawling about the surface of the balloon. This points up an important property of the primoridal fireball: at any point in space we should see the radiation coming in from all directions. One should not think of the fireball as originating in some sort of localized explosion. Consistent with the assumption of uniformity, the fireball radiation always would have uniformly filled all of space.

We also see from the model that, if the ants refrained from reproducing, the average number of ants per unit volume would decrease as the balloon is blown up. In a like manner the radiation can be regarded as a collection of massless particles—photons—and we see that the number density of photons decreases as the universe expands.

Finally, as the universe expands, the wavelength of the radiation increases. If the light were radiated in a distant galaxy, we would understand this as a Doppler shift: the light is shifted toward the red end of the spectrum because the galaxy is moving away from us. This means that the wavelength of the light as we observe it is greater than the wavelength as it was radiated in the distant galaxy. However, this behavior cannot be peculiar to light which originated in the galaxy. Thus we can see that if a photon of the fireball radiation happened to pass by the distant galaxy, an observer there could measure its wavelength, and an observation of the photon when it reached us would reveal that the wavelength has increased by just the same factor as is the case for radiation actually emitted from the distant galaxy.

It is seen from these arguments that as the fireball radiation propagates through space in all directions the universe expands, and the wavelength of the radiation increases, in effect stretched out by the expansion of the universe. Also, the density of the radiation photons decreases. The net result is that the characteristic blackbody shape of the spectrum is preserved, but the temperature of the radiation decreases. This distinctive blackbody spectrum of the radiation is an important feature because it would distinguish the primordial fireball from radiation produced later, in galaxies.

Again, these theoretical speculations came before the observations. The first experimental evidence possibly indicating a primordial fireball was found about 1965, with the discovery of an unexpectedly high intensity of extraterrestrial radio radiation incident on the Earth at a wavelength of 7 centimeters. Extrapolating from the known radiation from the Galaxy at much longer wavelengths, one would have expected that the radiation at 7 centimeters would be almost a hundred times smaller than the observed value. This radiation had of course been received in earlier instruments, but it was not identified because one could not distinguish it from noise originating in the receiver, or noise originating in the Earth which managed to get into the antenna. Only when these local sources of noise in the instrument were made small enough, and were well enough understood, could people realize that there was an anomalously high extraterrestrial background.

Is this new radiation the primordial fireball? To test this idea it is necessary to measure the spectrum of the radiation and see if it has the required blackbody shape. There are now observations of the radiation at four different wavelengths, in the range from 20 centimeters to 2.6 millimeters; the measurements fit the blackbody radiation curve so that the results so far are very encouraging for this point of view. It should not be too long before measurements at still shorter wavelength complete the test. If the blackbody spectrum of the radiation is verified, it will provide direct evidence that the universe has expanded uniformly away from a highly contracted state. This would be a remarkable confirmation of the big bang picture of the universe.

Accepting the general features of the big bang theory as valid, we are confronted by an interesting question as to whether the universe is open or closed, whether the universe expands indefinitely or else stops expanding and contracts back. A possible test of this is to look for the expected deviations from Hubble's law for very distant galaxies. The difficulty with this test is that when we look at very distant galaxies we see them as they were in the past, because of the finite propagation velocity of light. Because we do not know how the properties of galaxies are changing with time, it would be difficult to interpret an anomaly in Hubble's law for very distant galaxies: we would not know whether the anomaly is associated with the cosmological model or simply with the changing properties of galaxies.

A second possible test is to look for the expected deviations in the counts of galaxies to various limiting brightnesses. This test was attempted by Hubble, but it was unsuccessful because systematic errors in the measurement of the brightness of the galaxies became more and more important as he counted to dimmer galaxies, thus obscuring the effect he was looking for.

The newly discovered quasi-stellar objects provide a powerful new probe of the distant reaches of the universe. Whatever the nature of these objects, they appear to be much brighter than galaxies, so that they can be distinguished at much greater distances. This means that the expected discrepancies in the above two tests would be larger than is the case for galaxies, but again the results will have to be interpreted with caution: Can we be sure that the quasi-stellar objects existing in these early epochs did not tend to be systematically brighter or dimmer than the quasi-stellar objects we see about us now?

A third test, which is beginning to seem promising, is to measure the mean density of mass in the universe. If this mass density is less than a critical value which is determined by the known rate of expansion of the universe, then the universe is open; if the mass density is greater than this critical value, the universe is closed. The mass density due to galaxies of ordinary size can be estimated with some confidence, and it appears clearly too small, by a factor of about thirty, to make the universe closed. The difficulty with this result is that it neglects the possibility of relatively large amounts of dark or dim matter in the vast spaces between the galaxies. With the recent very rapid development of radio astronomy, and of observations from rockets and satellites above the obscuring atmosphere, it has become possible to look for some reasonably possible forms of dark material. For example, measurements of the electromagnetic radiation background above the atmosphere permit us to conclude that electromagnetic radiation is an insignificant contribution to the total mass density of the universe. Also, it has been shown that there cannot be very appreciable amounts of hot ionized hydrogen filling intergalactic space, for such a gas would produce a greater flux of X-rays above the atmosphere than is observed.

These observations do not yet rule out a relatively cool ionized gas. But if it were too cool, the hydrogen would recombine, and apparently intergalactic recombined hydrogen can be ruled out on the basis of the spectrum of the very distant quasi-stellar objects. If recombined hydrogen were present, it should have strongly absorbed radiation from the very blue side of the spectrum, and this does not happen.

There are forms of dark matter which remain important possibilities, and, if we allowed our ingenuity free rein, we could find forms, like intergalactic rocks of moderate size, which would defy any attempt at detection. However, if we confine attention to possibilities which are reasonably consistent with our cosmology, the list is not all that long. For example, we do not expect to find intergalactic rocks because rocks would have been decomposed in the original very hot fireball. Thus it may be that the great improvement in observations possible, now that instruments can be taken above the atmosphere, will permit a reasonable resolution to the problem of the mass density in the universe.

This question, whether the universe is open or closed, is of considerable philosophical interest because it tells us how the universe ends (if the theory is to be believed). It will be recalled that the open universe expands indefinitely, so that in time the stars burn out and all activity in the universe dies away. On the other hand, the closed universe stops expanding eventually and collapses back in on itself. During this collapse the radiation in the universe heats up again, eventually decomposes material, and reduces it to ionized hydrogen, as it was initially. The model also implies that the collapse goes all the way, to a state of unlimited high temperature and density. Until recently it has been possible to cling to the pious hope that this is a peculiarity of the over-idealized assumptions of exact uniformity. Recent theoretical work seems to prove, on the contrary, that such a collapsing universe inevitably develops a singularity, which is to say that the equations become meaningless. What the universe would do at this point is by no means clear.

Despite these puzzles there seems to be good reason to believe the general picture I sketched earlier: that the universe we can see about us is on the whole uniform and uniformly expanding. Beyond this, the conclusions to be drawn depend very much on personal tastes and preferences. I have drawn freely on my own tastes and preferences in presenting this description of cosmology, and will continue to do so in the following summary of some of the most interesting problems of the subject.

An important problem is presented by the observed tendency for matter to appear in the lumps called galaxies. This is a question of immediate interest: we see that galaxies exist, and we do not have to trace the expansion of the universe back very far in time before the universe would have been too highly contracted to contain galaxies, at least as they are today. This means that the galaxies apparently formed at an epoch when the universe was not in a very mysterious state, and therefore when the equations of physics are well behaved, and we would hope that a sufficiently perceptive application of these equations would lead to a reasonable explanation for the galaxies.

Probably a much deeper puzzle is the remarkable uniformity of the universe: Why did the universe start expanding with such a uniform mass distribution and such a highly regular, initial velocity distribution? This is not a consequence of General Relativity: the theory admits equally the regular, uniform universes and highly irregular ones. Our own preference for the uniform and regular universes comes in part from a feeling that regularity is more elegant than irregularity—this sometimes is stated as the cosmological principle—and in part from the much greater mathematical simplicity of the uniform models. It is truly surprising that nature has chosen a universe consistent with our own notions of simplicity, and I find little comfort in the statement that this is a result of the cosmological principle.

We do not know how to trace the equations of physics back far enough in time to find out what the universe was like before the big bang. One idea suggests that, before the big bang, the universe was collapsing in from a previous phase of expansion followed by contraction. In this way we could imagine a cyclic universe, repeatedly expanding, collapsing, bouncing, and again expanding. This has the advantage that the universe is eternal, in a sense, so that it places the problem of creation back in the infinite past, where perhaps it belongs. A minor difficulty is that the equations do not predict the necessary bounce from collapsing to expanding phase—in fact, the equations break down at this point. On the other hand, this is not very disturbing because it is more than likely that the presently known physical theory would be quite inadequate to describe the conditions at the time of the assumed bounce.

We do not know how the universe ends, whether it keeps on expanding forever or whether it collapses back in on itself. In the latter eventuality we are completely uncertain about the ultimate results of the collapse.

The list of problems is long, and of course we have no idea what will be uncovered in the search for answers. But this still is a living science, and we can enjoy the excitement of the search.

V

LIFE

• 28

Chemical Evolution of Life on Earth

MELVIN CALVIN

"Chemical evolution" refers to that period of the evolutionary history of the Earth during which the chemical components on its surface were changed from their primeval forms into chemicals from which living organisms could evolve and could develop. This last phrase, "from which living organisms could develop," is perhaps the principal connecting link between the purely scientific aspect of what I am about to say and the humanistic values which it may or may not bespeak.

Throughout history, man has repeatedly made efforts to discover something of his origin, in order to gain from those beginnings some understanding of his destiny. Man devises questions couched in the language of the particular era and of the particular subculture in which he lives; the answers are limited to the language of the question. To the question of origin, I propose to seek answers only within the context of the scientific and technical society in which we are now living. This is not to exclude other possible answers or modes of response to this question, but only to seek to provide some kind of base which is subject to a scientific, or technical, test.

The idea that living organisms appeared as a natural development in the course of the chemical transformation of the surface of the Earth is not new. Charles Darwin himself recognized that the basic notions of evolution which he formulated applied, in fact, continuously, not only throughout the appearance of living organisms and the development of their enormous variety, but back through the stages of history into the period which preceded the existence of living organisms on the surface of the Earth. This was recognized by him in a very famous observation which I think is worth repeating because it reveals Darwin's chemical concepts as held as early as 1874: "You expressed quite correctly my views when you said that I had intentionally left the question of the Origin of Life uncanvassed as being altogether *ultra vires* in the present state of our knowledge, and that I dealt only with the manner of succession. I have met with no evidence that seems in the least trustworthy in favour of so-called Spontaneous Generation. I believe that I have somewhere said (but cannot find the passage) that the principle of continuity renders it probable that the principle of life will hereafter be shown to be a part, or consequence, of some general laws. . . ." [1]

It was possible to find the reference to which Darwin referred. It was in a letter which he wrote much earlier, in 1871: "It is often said that all the conditions for the first production of a living organism are now present which could ever have been present. But if (and oh! what a big if!) we could conceive in some warm little pond, with all sorts of ammonia and phosphoric acid salts, light, heat, electricity, etc., present, that a protein compound was chemically formed ready to undergo still more complex changes, at the present day such matter would be instantly devoured or absorbed, which would not have been the case before living creatures were formed." [2] Darwin here exhibited two qualities: First, a remarkable perspicacity about the nature of chemistry and, second, an altogether characteristic conservatism about how much he knew and how much chemists knew at that time about the nature of molecules. In those days so little was known about the nature of molecules and their interactions and behavior that it was fruitless for him, and others like him, even to try to reconstruct the chemical evolutionary history of prebiotic times.

---

[1] *Notes and Records of the Royal Society of London,* Vol. 14, No. 1 (1959).
[2] *Ibid.*

The situation today, however, is different. We not only have a much more profound and intimate knowledge of the molecular constitution of living organisms and of how they function, but we also have a much more detailed and intimate knowledge of the fossil record as it exists in the rocks of the surface of the Earth. We have now two kinds of knowledge in much more intimate detail than Darwin had, and I think, therefore, we are justified in undertaking such a quest for the reconstruction of the chemical events which gave rise to living organisms insofar as we can do so.

ORGANIC GEOCHEMISTRY

It is relatively easy to recognize the hard parts of animals, and even of plants, after they have fallen to the sea or lake bottom and have been covered by the mud there; they can leave their impression in the rock formed from that mud or sand over many millions of years. And these morphological structures can be recognized for what they are. This, of course, is the basis for the study of paleontology. Such a fossil record can be traced back quite unambiguously for at least 600 million years. In fact, quite recently it has been reported that well-developed fossil forms of higher animals have been found in the northern reaches of the European continent, and of the Canadian part of the North American continent, which have been dated as early as 700 or 750 millions of years. But beyond that, the fossil record becomes less unambiguous.

In recent years another kind of fossil record has been recognized, a microscopic one which can be seen only with the assistance of new techniques for the examination of thin slices of rock—using both the ordinary microscope and the electron microscope. It appears that there are shapes, or forms, which may very well be the residues of primitive microorganisms (bacteria or algae) much older than 700 million years. Even more recently, it has become an accepted notion that such microfossils are indeed the residues of microorganisms which are at least 1 billion years old. Currently, there are suggestions that such microfossils have been found in rocks that are even twice that age, perhaps 2 billion years old.

It seemed to us some years ago, before any microfossils had been unambiguously recognized, that some other way of tracing biological history in the rocks should be possible rather than to depend on undisturbed macroscopic shapes that could be distinguished as animal or plant. The rocks, after all, have been crushed and compressed and heated and distorted, and one might expect such macroscopic shapes to be destroyed. However, some of the organic molecules which were part of the constitution of those living things might be expected to remain. We know now that many of them do remain.

The organic constituents of the fossil fuels, coal and petroleum, are the molecular residues of living organisms. Therefore, we thought it would be useful to explore the molecular constitution of ancient rocks even though they had been so metamorphosed with pressure and distortion that there were no visible fossils in them. We undertook to examine the very ancient rocks, which do not have any large amount of liquid petroleum in them. The amounts of hydrocarbons which these rocks contain are trivial (merely traces), and we have to use methods of analysis which will allow us not only to say there are hydrocarbons present but to say which particular hydrocarbon structures are there. This required a very special analytical technique, with very special instruments, which have only become available to us in the last half-dozen years.

To see if we could identify a certain kind of hydrocarbon molecule as characteristic of the residues of living things, we began by analyzing a relatively young rock which we knew contained the residue of living organisms. We took a rock whose geological origin was not in doubt; this was the Green River Shale which underlies most of the middle of the United States and is only 60 million years old. We were seeking to analyze the organic constituents of rocks of increasing age until we should come to a time when the organic constituents perhaps cannot be called the *residues* of living things but rather the *precursors* of living things. There should be some interval in time, as we go back, when the character of the molecules is changed. Whether it is a sharp change or a slow change is yet to be determined, but this is the interface in time which we are seeking. We do not know that we have yet found it. We have, however, found molecules which, we are fairly confident, are not precursors to living things, but are residues of living things.

Before continuing further, it might be worthwhile to say something about the time scale with which we are dealing and into which we must fit. We must have some idea of the age of the

Earth in order to do that: and the time of the genesis of the Earth is about 4,700 million years. Somewhere in this period, between 2 and 4 billion years ago, organic evolution must have begun. It is this asymptotic point which we are seeking in our chemical analysis of the rocks.

The first analytical result on the Green River Shale came by way of a physical separation, one from another, of the hydrocarbons contained in it.[3] There are many different substances present in the Green River Shale, all of which are hydrocarbons. We proceeded to make a crude separation between those which are composed of long straight chains of carbon atoms (with nothing but hydrogen atoms on them) and the others. This second group contains not only long straight chains but occasionally some branches. We can then separate these once more in a very special way. Even among the straight chains in this 60-million-year-old rock, the distribution of hydrocarbon varieties is not uniform. There are some dominant ones, such as those containing 17 carbons in a chain, as well as those containing 27- and 29-carbon-atom chains. This fact in itself is an important piece of information.

In addition to the straight-chain hydrocarbons, there appear chains which have branches in them. In this class there is a special one in which a single carbon atom branches from every fourth one in the chain. Such molecules are called isoprenoids and are related to the familiar substance rubber. In this group there is, in addition to a chain containing 20 carbon atoms (phytane), both a C_{16} isoprenoid and a C_{18} isoprenoid, as well as a C_{19} isoprenoid (pristane). The phytane has a special kind of molecular architecture which is not accidental or random, and we want to know where the phytane and also the pristane in this 60-million-year-old rock come from. We believe that they come from living things which contain the chlorophyll molecule in them. It has been suggested by J. G. Bendoraitis that the phytane and

---

[3]Gas-liquid chromatography was the method of separating the hydrocarbons. It is done by carefully cleaning the rock, grinding it up, and then carefully washing and extracting it. From that point it is no longer exposed to the open air lest it be contaminated. Next, it is ground to a fine powder, and the fine powder is extracted with a pure solvent. The solvent is passed through a long column and the various compounds which are contained in that solvent pass through the column at different rates, the smaller molecules going more rapidly and the larger ones going slowly.

the pristane come from the long hydrocarbon chain which is attached to the green material (chlorophyll) which is present in all green plants. If this is so, then there must have been green plants 60 million years ago, and this is the kind of evidence from which such a conclusion can be drawn. The steranes and triterpanes in the Green River Shale are more complex molecules, but they are related in origin to the simpler ones, such as the isoprenoids. The analysis of the hydrocarbon content of this 60-million-year-old rocks shows a very characteristic distribution and contains architectural structures which are easily recognizable as the products of living things.

To test this hypothesis, it is possible to analyze chlorophyll in the laboratory. Chlorophyll upon hydrolysis gives the carbon chain phytol with 20 carbons and four branches. By reduction, oxidation, and decarboxylation, a C_{19} carbon-containing compound is formed. By hydrolysis, hydrogenation, dehydration, and hydrogenation a C_{20} hydrocarbon (phytane) is made which we think comes from chlorophyll. The same route which is used in synthesizing phytol is also used in making pristane. This is the kind of evidence which indicates that living things were present when such molecules as phytane or pristane are found present in the ancient rocks. Mass spectrometric analysis of the steranes, which are the carbon skeletons of such familiar molecules as cholesterol and the sex hormones, in this ancient shale shows a very complex architecture, and the molecules appear to be related to the simple phytane and pristane. The steranes can also be created in the laboratory by the same kind of biological synthetic mechanism as pristane and phytane, but the mechanism is more complex.

Not only have we found these kinds of "molecular fossils" (compounds of specific chemical architecture) in a 60-million-year-old rock, but we have found them also in the Nonesuch Shale, which is 1 billion years old. We then went to still older rock, the Soudan Shale, which is dated by radioactivity and other methods at about 2.2 billion years. In the analysis of the Soudan Shale we found the C_{18} isoprenoids (pristane and phytane) present as well as another fraction which contains the steranes, a complex ring system.

Remember that the present age of the Earth is only 4.7 billion years. We have evidence here that there were living things capable of making complex architectural constructions already present on the surface of the Earth 2.5 billion years ago. Therefore, not much time is left to create these things which can perform such complex operations, and we now either have to lengthen the life of the Earth or devise more rapid methods of evolution.

Let us look at further evidence. When hydrocarbons are made by a non-biological method, that is, by passing a spark through one-carbon compounds, no single compound is built up to any great extent. The compounds are all built up about evenly, and there are no sharp peaks. Thus, even without knowing what the compounds are, the fact that we have very sharply defined distribution of compounds in the geological material is in itself evidence that the existence of these compounds is not the result of a random process of synthesis but rather something very specific and, therefore, non-thermodynamic. Further, the process must involve some information transfer, which can only mean the interposition of a living thing. The various kinds of architectures and reactions which we have found in the analyses of very ancient rocks can be reproduced by the living organism in the laboratory in a very complex sequence of events.

Thus far in our analysis we have reached back in time to rocks at an age of some 2.5 billion years and have seen no evidence that we have reached a period in which there are no living things. The molecules that we see at 2.5 billion years are the same as the ones we see in the younger rock of 60 million years. We have not yet reached back to the point of chemical evolution which must have preceded the present-day kinds of life. The oldest known sedimentary rocks of which I am aware (although some geologists and geochemists in the world may have better information) are about 3.3 billion years old and are found in South Africa. We have recently obtained a small sample of these very ancient rocks, and they do not contain very much organic matter in them. So far as I know, this is the oldest known rock in which we can hope to find organic matter. Obviously, there are still older rocks, since the age of the earth is 4.7 billion years, but of these rocks I have no direct knowledge.

PREBIOTIC CHEMISTRY

We will now approach the problem of chemical evolution from the other end. The previous discussion concerned the "historical" way, in which we went back in time to try to find when the geological record changes from biological to non-biological material; as yet we have not found that interface in time. We could start at the other end, that is, with the Earth in its primitive form 4.7 billion years ago, try to determine what the nature of its atmosphere was, and then examine the kinds of chemical changes which can be induced in such an atmosphere by the incident energy from the Sun or from cosmic rays, or from the Earth's radioactivity, or from the Sun indirectly through turbulence of the atmosphere and electrical discharge. All these are high-energy sources of radiation which will tear apart the simple molecules which are believed to be the primitive molecules of the Earth's atmosphere.

Such experiments can be done (and have been done) in the laboratory. We began those experiments in 1950, and they have been done now broadly all over the world in various ways. The one that created the biggest excitement was that of Stanley Miller in 1955, when he put ammonia into such a reaction mixture containing the primitive molecules. He obtained more interesting molecules, compounds which contain nitrogen atoms as well as carbon and hydrogen atoms. The presence of the nitrogen atoms in the reaction mixture gives rise to the amino acids which are the building blocks of proteins, and this, in turn, gives rise to a whole new kind of evolutionary chemistry. Miller proposed that the primeval compounds included water, carbon monoxide, carbon dioxide, methane, hydrogen, and ammonia. From these, in early reactions, other compounds were created which led to direct precursors of organic molecules in present-day living organisms.

There has been a great deal of work demonstrating the transformation from the primeval to the primitive molecules, but this is a long way from the macromolecules—the proteins, nucleic acids, and polysaccharides—which are the essential constituents of today's living organisms. How can we go from the primitive molecules to the larger ones?

These macromolecules of biology, which we know popularly as proteins, fats, and sugars, are all made in the same basic way and from the same primitive molecules. The macromolecules are made by a dehydration process which eliminates water in various ways and allows the molecules to recombine into long chains of amino acids, which are the proteins of living organisms, or, by similar reactions, into the lipids or polysaccharides. These are the molecules which are essential for the storage, transfer, and transformation of genetic information in living organisms; perhaps even intellectual information is stored and transferred by this same mechanism.

Just how do we get from the simple molecules that we started with in the first instance to the biopolymers? In each case we must remove water, so a way must be devised for removing water from the small molecules, hooking them together. Surprisingly, this removal of water must be accomplished while they are still dissolved in water. The animal and plant organism of today is performing this kind of operation constantly. However, it takes great ingenuity to duplicate in the laboratory what nature seems to do so efficiently.

In the first molecules formed from the primitive atmosphere of the Earth there was present a molecule which held its atoms of carbon and nitrogen together with three bonds rather than only one. It is called dicyanamide, and its triple bond between the carbon and the nitrogen provides a way of storing in chemical form the radiation energy which created the molecule. This triple bond can be used to take water out of the other molecules; we have tried it, and it works. Peptides, which are small fragments of protein molecules, and other more complex compounds can be made by taking the dilute aqueous solution of dicyanamide and simple amino acids and allowing them to react for a few minutes under the proper conditions to create polypeptides. We can take the water from two molecules (one molecule of the acid and one molecule of the amine) and make the dipeptide, and that water molecule has been attached to what was a triple bond of the dicyanamide. Polynucleotides, small fragments of the genetic material, are also created by this same mechanism, but the actual routes are not yet established.

We now have a chemical system, tested in the laboratory, which can accomplish the polymerization, that is, the putting together of simple molecules into chains on the way to protein. Even when we have made a protein, we are still a long way from a living organism, but we are getting closer. The protein, with some particular form of amino acid, will assume a secondary structure, that is, take a higher level of organization which is built in to the primary sequence of the chain itself; it is possible to get an enormous variety of molecules by arranging the twenty different atomic groups (one for each different amino acid) in different ways.

Once we get such a biopolymer, it is not just a loose piece of string in water under physiological conditions; rather, it takes a secondary shape that is part of the structure of the molecule itself. I am representing the long chain as a random coil, and when the conditions of the aqueous solution are suitably adjusted to something resembling physiological conditions, this random coil makes itself into a helix. This is now known to be a perfectly reversible phenomenon for some amino acids and polypeptides. The helical structure is stable and built right into the sequence of amino acids, and is called a secondary structure of the protein. It is intrinsically the stable structure for the polypeptide.

The same type of phenomena occurs in the polynucleotides as in desoxyribonucleic acid, which is made up of the sugar phosphate chain with four types of bases attached. This linear array of bases can carry information which is transferred from one cell to another. It also has a secondary helical structure which is stable. At 22 degrees Celsius the helix structure is apparent; if the material is warmed, there is a change to a random coil, accompanied by a color change; and if the solution cools down again, the helical structure returns.

The polymer when in a suitable environment will pack together, and the third order of structure, the visible, will begin to emerge. We are now approaching the level of structures which are visible in functioning within the living cell. A fourth order of structure appears when the separate molecules of collagen (a protein), in a suitably adjusted salt solution, aggregate and form quite visible collagen fibrils which are indistinguishable from the natural collagen fibrils which may be extracted from a living thing. We have put together such "artificial" fibrils, and they show the same visible structure as one sees in living things.

Still we have not arrived at the living cell. We have yet to make a membrane or a cell wall which will enclose the macromolecular structures. We are only beginning to get some understanding of the construction of the internal cell organelles of a living cell. In a living cell we have seen quantasomes, which appear to be the ultimate unit for the conversion of solar energy, and once we understand the construction of one of these quantasome-type units we will be able to reconstruct it from component parts. There are still other more complex cellular organizations, but none of these structures is yet accessible to us in terms of reconstruction. I think that as we learn of what they are constructed we will be able to reconstruct them in the laboratory, at least functionally, just as we have been able to reconstruct the protein molecules. As late as ten years ago the reconstruction of these protein molecules was considered to be outside the range of the chemist, but they no longer are; we can perform this reconstruction today. I think the cell organelles, such as the quantasome, will also come within the range of construction, but how soon that will be, I do not know.

If this evolutionary process is as sequential as it appears to us now, given a particular starting point of methane–ammonia–water–hydrogen, then presumably the evolutionary process will take place wherever that same condition occurs. As of today, it does not seem that any such condition has occurred anywhere else within our solar system. There was much speculation and even some argument that such a condition was not only possible but likely on the planet Mars, but the pictures which came back from that planet seem to indicate that the surface of Mars is not undergoing a weathering process like the surface of the Earth, that it has only a very thin atmosphere today (roughly 1 percent of the Earth's atmosphere), and that its surface has not changed in about 300 million years. It is therefore clear that Mars is not undergoing the kind of evolution that the Earth is undergoing. This in itself does not necessarily mean that some of the chemical processes which we now know must have occurred very early in the Earth's history (before 300 million years ago) could not have occurred on Mars at that time. In fact, it seems likely. We are looking forward to the time when we will be able to send to the surface of Mars what we call an automated biological laboratory which will land, take rock samples, and perform the same types of analyses of them which we are performing in the laboratory on the ancient rocks from the Earth. It is quite possible to miniaturize these analytical procedures and automate them so that the equipment can get to Mars, make the analyses, and telemeter the results back to Earth.

Long before that time, however, it will be possible for us to find out whether there is any kind of organic matter on the Moon. The Moon is not such an inviting place, since there are great extremes of temperature and less atmosphere than on Mars, and it is not likely that there will be anything alive there. However, the Moon is like a cold storage place. We get dust and meteorites on the Earth constantly, and so does the Moon. On the Earth these things are transformed (they are weathered, microorganisms eat them, and they are changed). However, on the Moon these objects will be there, unchanged, under the moondust. Perhaps in three or four years the astronauts will come back with lunar samples for us to analyze in our laboratories, and we earthlings will have one more opportunity to look back into time, toward our beginnings.

• 29

Life on Other Planets

COLIN S. PITTENDRIGH

No matter how much we may marvel at the immense diversity of life on Earth—past and present—biology remains restricted to but one display: splendid and complex though it is, it remains, in a sense, a single case. The prospect of examining a second case, perhaps on Mars, is at the root of the biologist's interest in life elsewhere, an interest that is profoundly significant to biology and to man.

We cannot fully know the precise course of the Earth's early chemical evolution, and the degree of contingency involved in the subsequent transition to a living organization of molecules; and for these reasons we cannot fully assess just how probable or improbable life's origin was at the outset of our own planet's evolution. Nor can we estimate to what extent the emerging picture of a single chemical basis of life on Earth reflects a physical necessity for living organization as against a mixture of physical sufficiency and historical accident. Can the catalysis essential to biochemical organization be effected only by proteins containing the twenty amino acids we encounter in cells? Are the nucleic acids the only polymers, for physical reasons, that can carry molecular information satisfactorily? Or are these and other empirical generalizations about life on Earth, such as optical activity, merely reflections of the historical contingency that gave such molecules first access to living organization, thus pre-empting the field and precluding realization of other physically sufficient molecular foundations for life?

To the extent that we cannot answer these questions, we lack a true theoretical biology as against an elaborate natural history of life on this planet. We cannot prejudge the likelihood of life's appearance on Earth; therefore we cannot confidently take the great inductive step when we are told by astronomers that there may be 10^{20} planetary systems elsewhere in the universe with histories comparable to our own. One thing is clear: if life is unique to our planet the probability of its origin must be almost unimaginably low. If, on the other hand, the probability is at all appreciable, life must be abundant in the 10^{20} planetary systems that fill the sky.

At stake in this uncertainty is nothing less than knowledge of our place in nature. It is the major reason why the sudden opportunity to explore a neighboring planet for life is so immensely important.

The biologist's interest in planetary exploration derives from Darwin. It is an extension of the evolutionary scheme he imposed on all biological thought about a hundred years ago. The general thesis that all contemporary life is an evolved modification of earlier life brings with it a powerful explanation of the curious unity that underlies biological diversity. Darwin's own thought developed in a period when biological structure was known principally, if not exclusively, at the macro or anatomical level. He was impressed by the unity of anatomical organization that underlay the superficial diversity of such things as men, apes, horses, porpoises, and bats. And he recognized that unity as something inherited from a common ancestry.

The rapid development of biology in the twentieth century has extended our knowledge to the micro level of structure—of cells, their constituent organelles, and even their molecules. Unity again underlies diversity. In the epithelia of daffodils, the muscles of men, or in the unicellular flagellate, a common organization is found—a continuum of membranes connecting the outside boundary through the endoplasmic reticulum to the nuclear membrane, mitochondria, chromosomes, and so on. And at the molecular level man and flagellate are singularly monotonous: everywhere catalysis is effected by similar protein enzymes, energy transferred by adenosine triphosphate (ATP), everywhere ATP is regenerated by the same electron transport system of mitochondrial proteins, identical enzymatic co-factors like nicotinamide adenine dinucleotide (NAD) occur in cell after cell, and everywhere hereditary information is stored and replicated by a system of nucleic acids.

The limited patterns which Darwin could see binding genera, families, orders, classes, and phyla of organisms are themselves subsumed under a still more general pattern of cellular structure and chemical constitution that binds all life on this planet into a common family, descendants of a common ancestry.

Darwin clearly glimpsed the potential extension of his evolutionary ideas to embrace the origin of life from the non-living world, but it is to the Russian biologist A. I. Oparin we owe the most explicit formulation of this great concept. It is the final link in the chain of scientific ideas underlying a philosophic naturalism: the unitary treatment of all nature without recourse to supernatural or occult forces. Oparin set the framework within which considerable subsequent discussion—and recently experimentation—on the origin of life has taken place. The central fact is that the planets, including Earth, have themselves undergone considerable physical change, development, or evolution of a sort, since their origin as part of the solar system. The chemistry of the crust and atmosphere has changed; and as part of that history Oparin envisaged the evolution of chemical complexity: the synthesis of organic molecules without the intervention of living cells and, later, the origin of a living system built from these molecules. The origin of life is thus conceived as a chapter in the greater evolution of the planet's surface.

Evidence from several sources suggests that the primitive atmospheres of the Earth and other planets were chemically reducing. Methane (CH_4), ammonia (NH_3), water (H_2O), and perhaps cyanogen (C_2N_2) were the probable constituents. Oxygen was probably absent or very scarce. In its absence the small reduced carbon compounds were free to react, producing simple carbohydrates, amino acids, and other organic units (monomers) of the complex polymeric molecules characteristic of the cell. These reactions, all endergonic, were energized either by electric discharge in the atmosphere or perhaps by the short-wavelength ultraviolet light which then could penetrate to the Earth's surface in the absence of oxygen and ozone. The molecules so synthesized could never have been abundant but in the sterile conditions then prevailing could accumulate. (Darwin glimpsed this too: ". . . at the present day such matter would be instantly devoured or absorbed, which would not have been the case before living creatures were formed.") They could, moreover, be concentrated in one way or another, as by adsorption of clays. At any rate, there was an historical opportunity in the course of the Earth's early development in which sufficiently complex molecules had accumulated in local concentration to the point where a particular association or organization of them could occur "spontaneously" and was of such a nature that it could replicate itself from the molecular components in the surrounding milieu.

This primeval spontaneous generation, as Darwin called it, differs in no way from the spontaneous generation which the history of biology has been at pains to demonstrate does not now occur. Its occurrence primevally was contingent on the unique historical sequence of conditions prevailing then but not now: the reducing atmosphere, the availability of an energy source, and especially the sterility of bodies of water.

The first self-replicating systems of molecules must have been heterotrophic: they must have utilized, as building blocks and as a source of energy, the organic units previously synthesized abiologically. The earliest form of respiration—the oxidative degradation of complex molecules to liberate energy—must have been anaerobic: oxygen was present at most in very small amounts, due to the photodissociation of water by ultraviolet light. The history of the initial heterotrophic organisms cannot have been long: the meager reservoir of building blocks would soon be exhausted. Subsequent events must therefore have included the evolution of autotrophic systems, those with biochemical competence (1) to synthesize the building blocks themselves from small molecules in the environment and (2) to exploit radiant energy systematically for such syntheses. This secondary evolution of photosynthetic autotrophs produced the great bulk of the oxygen in the present-day atmosphere and thus created the opportunity for a more effective, aerobic form of respiration in which oxygen is the terminal hydrogen acceptor in the reactions that degrade molecules to liberate energy. To this day both anaerobic and aerobic respiratory mechanisms persist. It is a curious fact that the aerobic mechanism is universally isolated in distinct structural entities (the mitochondria) of cells. Photosynthesis too is generally effected only in distinct organelles (the chloroplasts) of the cells. Mitochondria and chloroplasts manifest some striking chemical similarities and are unique among the cell's extranuclear components in possessing their own DNA—their own chemical heredity. They *could* be historically related, evolved descendents of an autotrophic organism that invaded and acquired a symbiotic relationship with another anaerobic, heterotrophic cell. Be that as it may, the speculation suffices to illustrate the general probability that the organization of cells surely evolved piecemeal, exploiting historical opportunity in sequence as it arose, and quite probably involved the beneficial mutual association of previously separate minor organizations. The origin and early evolution of organization is thus fraught with more imponderable contingencies than the strictly chemical evolution it followed and exploited.

The living thing is made from the stuff of the non-living world. It involves no qualitative novelty, no *élan vital;* it differs from the non-living only in its complexity and organization. The organization of its molecular constituents confers on the system, as such, those properties we recognize as "life." Of these, the most fundamental and defining is the capacity to store and replicate the information that specifies the organization. In all life on this planet that information is encoded in the linear sequence of the four (or five) monomers (mononucleotides) from which the long polymeric nucleic acids are built.

The evolution of life, subsequent to its origin, has been driven by the forces Darwin identified: mutation and (natural) selection. Spontaneous variation in the sequence of monomers in the nucleic acids constitutes spontaneous variation in the organization (living organisms) they specify. After its origin, life has persisted only as the product of preceding life, as the result of reproduction. Organisms that reproduce themselves with differing success make differential contributions to future generations. The latter are largely the product of those ancestors which, in the prevailing conditions, were the more effective reproducers. The process of differential reproductive success *is* natural selection; an inescapable feature of life's most fundamental property, it ensures its continuous evolution, which has the overall character of maximizing the organism's fitness to the environment it exploits. The passage of time inexorably ensures the invasion by life of a nearly incredible diversity of specialized environments, enclaves where particular organizational variations (reducible to DNA variations) reproduce not only adequately but better than other variations.

This character of biological evolution is fundamentally different from those historical sequences that preceded and made it possible: it demands the attainment of life's defining feature, self-replicating organization. The preceding events were (1) the development of organic and macromolecules and (2) their organized association into a minimal living thing.

The totally speculative nature of the chemical evolution was removed in 1953 when Stanley Miller performed his classic experiment: the synthesis of amino acids by an electric discharge in a model system of the Earth's primitive atmosphere. Since then he and other workers applying energy sources to presumptive primitive environments have synthesized the whole gamut of major types of organic molecules from which cells are fabricated; the list now includes amino acids and their simple polymers, carbohydrates and fatty acids, purines and pyrimidines, nucleotides—including ATP—and even oligonucleotides. What was inductive conjecture for Oparin is now experimental fact: the great chemical complexity of its molecular constituents does not, in the last analysis, require the intervention of the cell itself; it demonstrably evolves from the chemical simplicity characteristic of the planet's birth.

The element in Oparin's inductions remaining unverified is the crucial step of the unguided (uninformed) origin of a minimal organization of such molecules capable of self-replication. There can be no reasonable doubt the induction is valid, but there remains great uncertainty on how probable or improbable was that step between the prior chemical and the subsequent biological evolution. And the uncertainty frustrates us when we learn there may be some 10^{20} planetary systems in the universe. But for the uncertainty we could not escape "the great extrapolation": the further induction, from the single terrestrial case, that life is a common feature of the physical universe, that, as on Earth, it has emerged repeatedly as the most complex form of matter exploiting the prior chemical evolution. It is hard to escape that inductive step even now; if it is wrong, the probability of a minimum organization arising must be almost unimaginably low.

It is surely unnecessary to labor how tempting the induction is: if valid, the biologists' horizons would, in principle, be extended indefinitely. The comparative method so powerful in elucidating Earth's life would be available to attack a range of major questions currently obscure. Is the chemical pattern monotonously present in Earth-bound life a physical necessity or merely physically sufficient? Is its ubiquity here a reflection of historical accident, a pattern imposed on local posterity by the one physically sufficient system which—arising first in a stochastic process—preempted the opportunities afforded by the evolving Earth? Can the information essential to life be stored and replicated by polymers other than the nucleic acids? Must, indeed, the information be stored digitally in a linear sequence of monomeric units? Is the carbon-water basis the only milieu in which adequate molecular complexity can be built? How have other living systems evolved? Which of the empirical generalizations about terrestrial life are true generalizations and which are general only for the local case? In sum, to what extent are we limited now to a local natural history as against a truly general, theoretical science of life?

And beyond the strictly scientific harvest there is the less easily defined but more widely grasped general philosophic concern so succinctly packed into T. H. Huxley's phrase: "Man's place in nature."

The development of space technology in the last decade offers the hope, at least, of testing, validating, and exploiting the great extrapolation. We can now seek empirical answers to questions about the conditions for and the probability of life's origin. The tools are at hand to explore our own solar system and thus to base our inductions on a sample greater than one planet. Were we, for instance, to find living organization on either Mars or Venus, the more Earth-like and fortunately the nearest of our neighbors, we would *know* that life was indeed a common endpoint of matter's evolving complexity: occurring twice in one system it must be abundant in the some 10^{20} systems that fill the sky.

What do we know about Venus and Mars to encourage or tempt our enthusiasm to explore them as biologists? Venus, surrounded by a dense atmosphere containing water, is currently believed to have a surface temperature so high (about 640 degrees Kelvin on the dark side, 750 degrees on the light) as to exclude the possibility of any system we would recognize as living. There is, however, some residual uncertainty about that temperature value, which is based on radio measurements at wavelengths longer than a centimeter and confirmed by the Mariner II microwave radiometer, so that we should not wholly exclude Venus from our concern, especially in view of its atmosphere and water content.

Mars, however, in spite of Mariner IV photographs, remains, on the whole, the more likely prospect. It is the nearest and in some respects the most Earth-like of the planets in the solar system. Its mass is about one tenth that of the Earth; its equatorial diameter is some 7,000 kilometers, about one half of that of the Earth. The year is long (687 days) but the length of its day is curiously similar to our own.

The planet has retained a very thin (some 5 to 10 millibars, possibly a little higher) atmosphere which we believe, principally on Mariner IV evidence, to be dominated by carbon dioxide. Oxygen is absent—or rather not certainly present as more than 0.1 percent (by volume) of the atmosphere. Water vapor has been confidently identified spectroscopically as a very minor constituent.

The thin anoxic atmosphere almost surely implies a heavy flux of ultraviolet light at the planet's surface. The surface temperatures vary widely with latitude, season, and time of day. The diurnal range overlaps that of Earth; at some latitudes and seasons it is about 100 degrees with a high of 130 degrees Celsius.

Prior to Mariner IV our knowledge of the surface was limited to the facts that there were two white polar caps and between them a mosaic of so-called dark and bright areas. The bright areas appear orange-ochre or buff and have traditionally been considered deserts. Some observers have described the dark areas as green, and they have long been a focus of interest in speculation about Martian life. The polar caps wax and wane seasonally; as the spring advances a wave of darkening proceeds through the dark areas toward and even beyond the equator. Other evidence of seasonal change in these dark regions comes from polarimetric studies which suggest the surface is covered with small, submillimeter particles; the curve on which this inference is based shows seasonal displacement in the dark but not in the bright areas.

The polar caps have been generally considered to be ice, or rather hoar frost, on the basis of spectroscopic evidence, but recent theoretical work, based on the inferences from Mariner IV that the atmosphere is nearly pure carbon dioxide, suggests that the caps contain or are covered by solid carbon dioxide.

The evidence has never indicated that present-day Mars had much surface water in the liquid phase, and the Mariner photographs imply that the scarcity is of long standing. The surface is pock-marked with craters. The slight erosion of their edges is most likely eolian but the role of agents other than wind cannot be fully excluded. Certainly there is no evidence that the surface of Mars has gone through a physiographic evolution comparable to Earth's: there is no evidence of ploughing by mountain-building and water erosion, no sedimentary formations. The absence of conspicuous orogeny goes with the failure of Mariner IV to detect any magnetic field attributable to Mars; its interior is probably "dead" and the planet as a whole has a more Moon- than Earth-like character.

What little we know of Mars is certainly the picture of an environment hostile to the earthbound like we know. It is, however, another thing to conclude, as some have done, that Mars is surely lifeless. Nothing we know precludes the presence of organisms: absence or near absence of oxygen is, of course, not crucial; the severe temperatures are not prohibitive; the ultraviolet flux we suspect at the surface can be shielded against and even exploited in conceivable ways; and water vapor though scarce is an identified atmospheric component. Water is surely the crucial item in the known list of marginal conditions. It remains likely, however, that the polar caps are largely water even if covered with carbon dioxide. They could well be a permafrost layer emerging at the surface in the polar regions, and to this extent we cannot exclude at their edges—even if subsurface—water maintained in the liquid phase by an adequate salt content.

The biologist is forewarned against the conclusion from existing "facts" that Mars is lifeless by familiarity with two aspects of the terrestrial case. First, the crude averages presented by current measurements of Mars provide no clue to what local diversity of microenvironments may exist. The average conditions of the Earth's deserts and mountain tops mask the existence of small niches where life flourishes. And, second, life itself is characterized by its evolutionary resourcefulness. Here on Earth exploited environments include hot springs close to the boiling point, the immense pressures of the ocean floor, saline pools in the polar caps, the pages of library books, bottles of sulfuric acid, and exposed rocks above 20,000 feet. No spoonful of sand from the Mohave desert is free of bacterial life. Which of these environments is the most hostile depends on the organism you ask; it depends on the monomer sequence of its deoxyribonucleic acid (DNA). Given the primary qualifications of information storage and replication, the combined forces of mutation and selection ensure the exploitation of life of every "marginal" environment. It is clear, of course, that any life existing on Mars will be, by virtue of the selection pattern that environment imposes, radically different from the more familiar forms on our own planet. Discussion of the pros and cons of Martian "humanoids" by laymen and even professional evolutionists is distressingly and incredibly beside the point as far as the real issues are concerned.

The adequacy of an environment to support life is not necessarily adequacy to provide its origin. But the fact that we cannot do this for Mars is not to the point; we could not have done so a priori for the Earth; in fact, our theoretical incompetence here is the primary motivation for an empirical study of other planets. The one thing we might confidently specify as a condition for the whole generating sequence is the existence initially of reducing conditions to promote the development of an organic chemistry. That premise for chemical evolution is in fact one significant element in our *conclusion* that conditions here were reducing. We still lack secure knowledge of how and why the early terrestrial atmosphere was reducing or even what its specific composition was. We cannot therefore usefully attack the hypothesis of Martian biopoesis, as one chemist has done on the grounds that the escape velocity of hydrogen on Mars is too low to assure an enduring reducing atmosphere. The plain fact is that there is nothing we *know* about Mars, its history, or the conditions essential to biopoesis that renders the hypothesis of Martian life and its local origin out of the question. Indeed, it takes some over-confidence in one's present understanding to assert that it is even unlikely.

In defending this position, I am not adopting the view that the biologist's interest in Mars and its exploration is contingent on the actual presence of life there. The existence of any organism—let alone ninety-nine of G. G. Simpson's "humanoids"—is not a prerequisite for the serious biologist's interests. To assume otherwise is to miss the central point. The evolutionary thought of Darwin and Oparin, and the experimental biochemistry of the last few decades, impels us to treat life as only one endpoint of a truly cosmic pattern of evolutionary processes generating complexity.

The biologist joins the physical scientist—astronomer, planetologist, and geochemist—in a common concern with the evolutionary processes of stars, planets, molecules, and molecular systems capable of replication. The biologist's real goal in planetary exploration is to enlarge his understanding of the origin of life, the conditions on which it depends, and its overall probability in nature. To do this he must enlarge his understanding of molecular and especially planetary evolution in general; and the exploration of our own solar system, now technically possible, is the surest and, on several issues, the only way to attain that understanding. We must expect and understand several patterns of chemical evolution, some congenial to life's origin, others not. What promotes the differences? When life is absent in the presence of a rich organic chemistry, why? How far can chemical complexity proceed in the absence of life? Sterile planets are thus of major interest in the fundamentally comparative study that must underlie any general evolutionary discussion of the planets and hence of life's origin.

• 30

Intelligent Life in Other Parts of the Universe

F. D. DRAKE

There is perhaps nothing so fascinating as the possibility that somewhere in the sky are civilizations which we could contact if we but manipulated the right instrument. Mankind has been tantalized by this thought ever since Galileo first turned his telescope to the heavens and found that other worlds traveled as we through the void of space. Indeed, if a worldwide election were held to determine what marvel we would most wish science to produce, communication with another civilization would rank very near the top. Thoughts of other civilizations sometimes reflect a wish for escape to a Utopia—to the "good life" many people assume, perhaps naïvely, other peoples would have achieved. But the serious and legitimate motivation behind our interest comes from the certainty that contact with another civilization would produce the greatest bonanza of scientific and historical facts of all time. Perhaps even more important, such contact would go far to answering some of those most personal questions we all ask ourselves at times: What is the significance of life in the universe? What does it mean to be a human being? What is my importance in the scheme of things?

The explosive growth of scientific and technical knowledge of recent decades has made feasible contact with another civilization. This required the achievement of two goals. One was the accumulation of compelling evidence that intelligent life is not rare in the universe, and the other the mastery of technology which could detect reasonable manifestations of intelligent life over the distances which separate the stars. By "reasonable manifestation" we mean a level no higher than already attained.

The high probability that there is intelligent life elsewhere in the universe is easily seen when we consider, first, that there are 100 million million million stars in the universe, of which the Sun is the most average, pedestrian example. There is nothing about our Sun which suggests that anything out of the ordinary has ever happened to it. Thus, it is reasonable to assume that the history of the Sun and the solar system has been repeated countless times in the history of the universe. Secondly, we now know enough of the chemistry of life to realize that, far from requiring some freak set of circumstances for life to develop, the conditions that existed in the early history of planets like the Earth would have made easy the development of life. As we shall see, the fossil record on Earth suggests that intelligent life will evolve often on life-bearing planets. Given this simple combination of facts, the existence of not only a few, but an enormous number of, civilizations in space seems assured. As Lee DuBridge, the president of the California Institute of Technology, has said, it is not the detection of life beyond the Earth which would be amazing: rather it would be startling if we failed to find it.

Technology has given us more than one practicable means of communicating across space. We have powerful radio telescopes, optical telescopes employing lasers, and perhaps in the future rockets which could reach the many hundreds of light-years which probably separate us from the nearest civilizations. We are offered perhaps the greatest opportunity in the history of mankind, the chance to join the community of civilizations of space with all the benefits—scientific, material, and philosophical—that accrue to members.

It was the development of these sophisticated communications systems over the last ten years which brought our attention to this opportunity and caused the scientific world to begin to turn its analytical skills to a study of how mankind might best exploit it. As we now see it, the problem can be divided into two parts. First, we must use astronomical and biochemical information to predict the number of technologically advanced civilizations now present in our Galaxy. From this number and our knowledge of the structure of the Galaxy we can predict the distance to the nearest civilizations. Once possessing an estimate of this distance, the second part of the problem is to choose the technological approach, be it rocketry, radio waves, or something else, which is most likely to detect a civilization over such a distance.

The number of technical civilizations depends equally on many things. It depends on the rate at which stars are being born in the Milky Way. From astronomical studies of the numbers of very young and very old stars we know with considerable accuracy that about one star per year is born in the Milky Way. How many of these will have planets? All modern theories of the formation of stars require that a second body or bodies be formed simultaneously so as to serve as a dumping ground for the spin or angular momentum possessed by the clouds which form stars. These clouds spin slowly as a result of their being members of our twirling Galaxy, the Milky Way.

In fact, when we study the stars we find that indeed about half are double stars, that in our own solar system 98 percent of the angular momentum is in the planetary system and not in the Sun, and that the distance from the Sun to the major planets of our system, Jupiter and Saturn, is about the same as the average distance separating the members of double-star systems. Thus, the theories and facts fit together and strongly suggest that the stars which appear to be alone in space are actually accompanied by planetary systems resembling ours. Nevertheless, all of this evidence is indirect, and we have so far observed but one planetary system, our own. Our confidence in our estimate of the number of planetary systems would be greatly increased if we could observe but one other planetary system in the vicinity of the Sun. This is perhaps a prime task for the great telescopes which will be orbited above the atmosphere of the Earth in the near future. Undisturbed by the turbulent atmosphere of the Earth, these telescopes may be able to detect the dim planets which should accompany the nearby stars.

Given a planetary system, only a few of the planets will be suitable abodes for life. To the best of our knowledge a planet may give rise to life only if the temperatures are somewhat between roughly the boiling and freezing points of water. In our solar system that seems to include three planets—Earth, Mars and Jupiter, the last of these having the appropriate temperatures only deep in its extensive atmosphere. Again, our theories predict that about the same proportion of planets will be suitable as abodes of life in other planetary systems. But will life arise? This subject has been discussed at length elsewhere in this book. The answer seems to be Yes. Given only the abundance of the chemical elements which exists throughout the universe and thereby in newly formed stars, the sources of energy which include lightning, ultraviolet light, and cosmic radiation and heat, and the billions upon billions of years over which Sun-like stars bathe their planets in a constant light, life will almost certainly arise. Here again we have observed but one example, terrestrial life. It is very important for spacecraft to validate our theories by studying the biochemistry of Mars and Jupiter.

Once given this life on any planet whose surface area is finite, and we surely know this will be true, the developing life will always in time encounter a shortage of food. With this comes the unrelenting competition between organisms which continues from the simplest one-celled organisms to the cry in the forest, and leads to the preservation and development of the superior creature. In the fossil record contained within the rocks of the Earth we have watched the succession over billions of years of one more-capable creature after another, each trying some new device to aid survival, be it camouflage, many legs, large size as in the dinosaurs, and so forth. Of all the things that have been tried by the creatures of Earth, only one characteristic has continuously been retained and improved throughout the entire succession of species—intelligence. Thus it appears that we could expect intelligence to evolve commonly.

Taking all of this into account, we find the abundance of life controlled principally by the rate of star formation and the percentage of single stars, and hence probable planetary systems in the sky. We deduce a surprising result: About one new intelligent civilization appears in the Milky Way a year.

But will all of those which have been developed in the history of our Galaxy now be detectable? The answer seems to be No. Only those will be detectable which today are still radiating into space detectable amounts of power. We have come to suspect that the longevity of technical civilizations in a condition releasing great amounts of power into space is limited. The number of detectable cvilizations is just proportional to the average longevity of civilizations. What limits this longevity? Perhaps cosmic accidents, perhaps the catastrophe of a nuclear war. Much more likely is that such civilizations become so sophisticated technologically that they are able to control power extremely well, and they no longer wastefully flood space with energy they do not use. They may, for example, transmit messages through tubes instead of through the atmosphere of their planets. When this happens we lose the ability to detect them simply because they are too advanced. It is also possible that the longevity is limited by loss of interest in technology. We do not know; in fact, perhaps the most interesting result that can come from the detection of extraterrestrial civilizations is a knowledge of this longevity and what typically determines it throughout the universe.

Unfortunately, our estimate of the number of detectable civilizations depends directly on estimates of this longevity. Numerically, in fact, the number of detectable civilizations in the Galaxy will equal the mean detectable longevity of civilizations in years. We have no information to go on except our own era of detectability, which we hope will be very much longer than the ten years or so it has now lasted. The estimates of longevity now made are based on no sound scientific data but rather on the estimator's opinion of the intellectual vitality of the human race. They range from ten years to a million years, with the most common estimate being one thousand to ten thousand years. Perhaps this longevity is lengthened by the actual act of civilizations communicating one with another. In any case, if we accept these estimates of the longevity, there are then about 10,000 detectable civilizations in the galaxy. We reach one very important milestone: the distance to the nearest detectable civilization is about one thousand light-years.

We are now ready to turn to the question of the engineering approach which is most likely to detect other civilizations. What criteria might we use to lead us logically to the means of interstellar discourse most commonly used in the Galaxy? Surely we cannot use as a guide the technological methods in which we are most proficient, such as radio communication, because other civilizations may have achieved their great successes in other areas, such as infrared communications. Nor can we rule out a possible mode of communication, such as nuclear-powered rockets, simply because we have not yet succeeded at it. In fact, if we are to base our decision as to the best means of communication on facts and principles that are not a consequence of the peculiarities of human technological history, we must use judgment criteria that will clearly be common to all civilizations. This means that we must confine our guidelines in choosing the most effective technology to the laws of physics and the arrangement of the universe, and their inevitable effects on the characteristics of living things.

Of the criteria which can be drawn from these sources, perhaps the most useful is the idea that economy or thrift will be practiced universally in interstellar communication. It may seem that economy or thrift is a peculiarity of mankind or of life on Earth, but in fact it is a principle practiced by all living things simply because the resources to support life are limited on all other planets as on the Earth. Thus the ability to practice economy with the available resources has enormous survival value and will be developed in all living things. Therefore, it is quite reasonable to believe that the concept of economy is well established and practiced in civilizations throughout the universe.

The recognition of the importance of economy leads to the rather surprising conclusion that rocketry or spaceships will not be the prime mode of interstellar communication. In the space age we have become accustomed to the idea that rockets are extremely effective ways of moving from one heavenly body to another. This is surely true within the solar system. However, when it comes to the vast distances which separate the stars, rockets suddenly become a very unexciting means of transportation. This derives from the need to travel hundreds to thousands of light-years, in turn requiring that the rocket travel at nearly the speed of light if it is to accomplish its task within a satisfactory time interval. But when we try to propel rockets at the speed of light, we find that the theory of relativity works against us and demands rockets of such a size that we not only do not know how to build them, but they would be outlandishly expensive. Even a nuclear-propelled rocket which travels at 99 percent of the velocity of light to another star and returns with a payload of one ton would weigh 100 million tons at takeoff. Such a rocket is preposterous, of course, and brings us to an important conclusion: the transmission of material across interstellar space will happen rarely if ever. The great distance between stars and the theory of relativity force us to send not matter, but only information, which is really what we want in the end.

In fact, there is an economical means to send information at the speed of light—electromagnetic radiation such as light, radio, and infrared waves, and X-rays. Even though we are young technically, the equipment we have built can already send a sixty-word telegram many tens of light-years for less than one dollar's worth of electrical energy. This is surely a bargain compared to the fantastic rockets required to achieve the same result. The laws of physics offer no other economic competitor. Thus it is electromagnetic radiation that will be most likely the interstellar messenger of civilizations in space.

But can we say more than this? After all, there is an enormous number of possible frequencies within the electromagnetic spectrum which might be used. Again, economics aids us in further narrowing the choices. The quantum theory of radiation tells us that electromagnetic radiation comes in units called *photons,* each of which contains an amount of energy which is proportionate to the frequency of the photon. This means that light photons, for example, will contain about a million times as much energy as a radio photon. Yet in a crude sense each can convey only the same amount of information. The cost of generating a given amount of energy is about the same everywhere in the electromagnetic spectrum. This means that it will cost about a million times as much to convey the same amount of information at light wavelengths as at radio wavelengths.

This then guides us economically to the radio wavelengths as the prime candidates for interstellar communication. But should we carry this argument further and deduce that we should use the lowest possible frequency? When we consider this possibility, we are confronted with the structure of our Galaxy, which places between the stars a vast sea of cosmic rays orbiting in magnetic fields and thereby producing intense radio emission on the lowest frequencies. This radio emission is captured by radio telescopes no matter where they look in the sky and appears as noise or static in the receiving system of the telescope. Thus this radiation from cosmic rays in our Galaxy acts to jam radio signals on the lowest frequencies and causes us to have to send many photons instead of the one photon required on higher frequencies to transmit an amount of information.

We can in fact study these two limitations mathematically—the cosmic radio noise on one hand and the photon energy effect on the other—and deduce a frequency at which information is transmitted with maximum economy. It turns out to be a frequency of about 3,000 megahertz, a frequency well above the present television bands and commonly used in radar applications. It is a frequency which penetrates the Earth's atmosphere and which is well received by many of the radio telescopes of the world. When we examine the telescopes we have for this frequency band, we find that the equipment now existing on Earth could detect a reasonable signal of the type that we ourselves transmit over a distance of 1,000 light-years or more. Thus the detection of civilizations at the distance at which we think they are located and on the more likely frequencies is possible now.

One preliminary search for extraterrestrial radio signals has been made. In 1960, the 85-foot telescope of the National Radio Astronomy Observatory at Green Bank, West Virginia, was used to look for radio signals from the two nearest and apparently single Sun-like stars. These are Tau Ceti and Epsilon Eridani, each about 11 light-years away. This was the search known as "Project Ozma." With each star, frequencies close to 1,420 megacycles per second were tested and no evidence for signals was found. This is the frequency radiated by hydrogen atoms which occupy space between the stars, and has been suggested as a frequency for interstellar communication because it is a unique natural frequency associated with the utmost abundant element, and all technical civilizations would know this.

The lack of success of this equipment confirms that the search will not be easy. Within 1,000 light-years there are about 10 million stars, only one of which may have a detectable civilization. Thus we must search 10 million stars, examining many frequencies since it is not guaranteed that the arguments we use to pick the best frequency will be the same ones made in other civilizations. In fact, in our search we should carefully guard against the possibility that they are not transmitting special messages to us on the "best" wavelength at all. After all, we do not blindly send out messages on the most economical wavelengths for the consumption of other civilizations. They, as we, may use radio signals only for their own purposes and not for the benefit of other civilizations in space. Since this may be the case, we should use a search strategy which will enable us to detect the signals a civilization uses for its own purposes. These may occur on any frequency. Fortunately, such search techniques have been developed. They utilize receivers which receive many frequencies simultaneously, store the measurements of the received radio energy in a computer, and make a sophisticated mathematical analysis of this information. Methods of analysis have been found which detect the ensemble of signals from a civilization despite the fact that no individual signal from the civilization is itself detectable. It is thus possible to establish that signals exist even though no individual signal can be received. Such techniques amount to *civilization detectors* and we should surely use them in our search if we are to have the highest chance of success.

If we plan to do this—and remember that 10 million stars may have to be tested—we find that such a thorough search will require the use of a very large telescope, say, 100 meters in diameter, a very complicated receiving system, and a large electronic computer for a time of at least thirty years. The total cost of the experiment over this time will add up to something like $60,000,000. Thus the time and monetary resources are formidable. Yet it is small when compared to the tasks we have undertaken in the exploration of our solar system. The possible fruits of this project—contacts with other civilizations in space—are so great that the project seems very worthwhile. There is a real hope that in our lifetimes such a search will be conducted successfully. The results could greatly enrich all human life.

Glossary of Technical Terms

Adams-Williamson equation. Expresses how density changes with pressure within the Earth. L. H. Adams and E. D. Williamson first used the equation in the 1920's to calculate density variations in the mantle.

Adenosine triphosphate. See *ATP.*

Adiabatic compression. One of a class of adiabatic processes, characterized by a change in matter without transfer of heat. For example, an ideal gas, perfectly insulated, would on adiabatic compression not only rise in temperature but some of the energy of compression would yield a pressure change greater than had compression been effected at constant temperature (by drawing heat away through a cooling system).

Aelosphere theory. Argues that solar energy can be converted into mechanical energy in the form of dust storms and that these, in turn, would generate heat by their frictional effects on the surface of a planet and would also serve to retain the heat, insulating the planet from space.

Aeronomy. A term coined about a decade ago by the distinguished English mathematician and geophysicist Sydney Chapman. It is the study of the physics and chemistry of the atmosphere, especially the upper atmosphere.

Albedo. The reflecting characteristic of an object, defined as the ratio of diffusely reflected to incident light. Because the albedos of known materials can be determined in the laboratory, comparisons with similar albedos of heavenly bodies can suggest something about their surface natures. Thus, the Moon's albedo is less than one-tenth, for it absorbs more than 90 per cent of incident radiation, which indicates that its surface materials are quite dark.

Alpha particles. Positively charged nuclear particles made up of two protons and two neutrons. The nucleus of helium is so constituted, and thus a helium atom stripped of its electrons is an alpha particle. Various nuclear reactions led to the emission of alpha particles from heavier elements. Fast-moving alpha particles are called alpha rays.

Amino acids. The principal constituents of proteins, from which they have been isolated. There are about thirty of them. S. L. Miller also synthesized them by subjecting a gas mixture of ammonia, hydrogen, methane, and water vapor to electric discharges.

Ångström. A unit of length named after the Swedish spectroscopist A. J. Ångström and used in expressing the wavelengths of light. It is 10^{-10} meter in length.

Angular momentum. The product of the moment of inertia of a rotating body and its angular velocity or rate of rotation. (See *Moment of inertia.*)

Anticyclones. See *Cyclones.*

Astronomical unit. The mean distance between the Sun and the Earth (some 149,600,000 kilometers or about 93,000,000 miles). It is largely used for distances within the solar system.

ATP. The commonly used abbreviation for adenosine triphosphate. It is a derivative of adenosine, which consists of carbon, hydrogen, nitrogen, and oxygen. Like adenosine, ATP occurs in muscle extract. It is important in sugar metabolism, which suggests its significant role in transferring phosphate-bond energy. ATP is produced in a complex photosynthetic reaction.

Autotrophs. Plants that can carry out photosynthesis, building up carbohydrates and proteins out of carbon dioxide and inorganic salts.

Basalt. A type of igneous rock, dark gray, black, or green-black in color and fine in grain. It is rich in calcium feldspar, poor in potassium feldspar, which characterizes granite, and often rich in olivine.

Black body radiation. The radiation that would be emitted by an ideal "black body," i.e., one which absorbs all incident radiation. The intensity and spectral distribution of black body radiation is a function of only the temperature of the black body. The formula for spectral distribution was derived by Planck from the quantum hypothesis (see *Planck's radiation law*). A black body can be approximated in practice by a nearly completely enclosed cavity, such that radiation emitted and absorbed by the internal walls is in equilibrium at a given temperature; the radiation is observed through a small hole in the cavity wall.

Bremsstrahlung. A German word meaning "braking radiation." It is used to describe the radiation produced when a charged particle is decelerated upon passage close to other charged particles. The radiation has a continuous spectrum. The term is most commonly applied to X-rays produced by electrons passing through matter.

Cepheid variables. Intrinsically variable stars whose luminosites fluctuate with periods of less than fifty days.

Chloroplasts. The parts of the cells of green plants capable of photosynthesis. They are the cell structures in which sugar synthesis occurs.

Chondritic (stony) meteorites. Those which contain basic minerals, feldspar, and nickel-iron.

Collagen. A fibrous, white, gelatin-like protein, good in tensile strength. Thin and long, collagen fibrils are the main supportive protein of skin, tendon, bone cartilage, and connective tissues.

Critical point. That point on a temperature-pressure diagram where two phases of state merge. For example, a liquid and its vapor, under certain temperature and pressure conditions, exist in a single state, where this means that the volumes of liquid and vapor are identical.

Cyclones. Large, low-pressure wind systems that usually bring foul weather and often great storms; they rotate as they pass over land and sea, counterclockwise in the Northern, clockwise in the Southern Hemisphere (as viewed from above). Anticyclones, on the other hand, are relatively high-pressure atmospheric systems, characterized by a minimum of cloudiness or storms; these "highs" rotate in a direction opposite to that of cyclones in each of the two hemispheres.

Desoxyribonucleic acid. See *DNA*.

Dipeptide. See *Peptides*.

Dipole. Literally meaning "two poles," this term denotes two opposite electric or magnetic charges that are separated by a small distance—e.g., a bar magnet or a polar molecule. By extension, it is also used to suggest the simplified picture of the magnetic field of the Earth as represented by a hypothetical bar magnet embedded in its interior and creating lines of force arching through space from one magnetic polar region to the other.

DNA. The common abbreviation for desoxyribonucleic acid. It is a long chain polymer. DNA and RNA are the two nucleic acids (see below). DNA is believed to be the transmitter of genetic information in biological systems.

Doppler effect. Refers to the apparent change in frequency or wavelength of waves caused by relative motion of the source and detector. Thus the pitch of a train whistle, as it moves toward a stationary listener, rises and then, as the train passes and leaves, falls. A Doppler shift is the difference between the observed frequency and the frequency of the same source at rest. The "red shift" in astronomy, usually interpreted as the Doppler shift, has attracted attention because of its revelance to our notions of the universe. If a star and the Earth are moving closer together, more waves are received in a given period, and this is revealed by a shift of the frequencies typical of the star's spectrum to a higher frequency or toward the violet. If the star is receding, fewer wavefronts arrive in a given period, and the frequency shifts lower to the red portion of the visible spectrum. The cosmologically interesting red shift corresponds to the apparent velocity of recession of remote galaxies.

Ecliptic. Refers to the plane of the Earth's orbit about the Sun and to the great circle cut on the celestial sphere by that plane. The celestial sphere is an imaginary sphere, an idealization of the sky against which the terrestrial observer sees the Sun, the planets, and the so-called fixed stars.

Eclogite. A kind of coarse-grained rock consisting of garnet and pyroxane.

Endergonic. Requiring the expenditure of energy to effect a biochemical reaction.

Endoplasmic. Refers to the inner part of the cytoplasm of a cell. Cytoplasm—protoplasm exclusive of the cell nucleus—consists of a thin, viscous outer layer (ectoplasm) and of the inner, watery, and granular endoplasm.

Epithelium. One of the four basic animal-body tissues. It is a cellular and membrane-like tissue that covers external body surfaces and lines internal vessels and other small cavities.

Galilean satellites of Jupiter. The four large ones (Io, Europa, Ganymede, and Callisto) were discovered by Galileo in 1610. The other eight are very much smaller and fainter.

Gamma rays. Electromagnetic radiation, similar to X-rays but shorter in wavelength and hence more energetic. More specifically, a gamma ray is a quantum of electromagnetic energy emitted by a nucleus under certain conditions.

Gravitation. A universal attraction between any two pieces of matter. As expressed by Newton, every particle of matter attracts every other particle with a force proportional to the product of their masses and to the inverse of the square of the distance between them. Gravity is a more restricted expression, referring to gravitational acceleration associated with a particular body, such as the Earth or a specific star. The constant of proportionality, alluded to above, is known as G, the gravitation constant, and has been measured. It should not be confused with g, which is the acceleration due to gravity of the Earth.

Hydrogenation. The process of combining hydrogen with another substance.

Hydrolysis. The alteration or decomposition of a substance by water.

Hydrosphere. The envelope of water covering the Earth's surface, including oceans, lakes, rivers, and ice sheets. It includes also the water vapor in the atmosphere.

Infrared. Electromagnetic radiation of wavelength shorter than microwaves but longer than light waves. The wavelength range runs from about 0.75 to 1000 microns. We experience infrared radiation as heat.

Ion. An atom or molecule which has lost or gained one or more electrons, making it charged, positively or negatively, and therefore electrically active. Ionization is the process of creating ions, common in liquids and gases. Thus, ionization takes place in the upper atmosphere where ultraviolet light and X-rays from the Sun dissociate atoms and molecules. The ions there, especially the free electrons, account for the electrically active nature of the ionosphere.

Isoprenoid. Pertaining to isoprene, a liquid hydrocarbon synthesized by distillation of some petroleum substances (e.g., naphtha) and rubber.

Isostasy. Literally, "equal standing." The term refers to a concept of balance of the topography of the Earth. Thus, high mountains and their roots beneath are less dense than the substrata of ocean basins, accounting therefore for a certain equilibrium between "highs" and "lows." The term *isostatic equilibrium* is common in geology, and refers to the hypothesis that columns of rock, with identical cross sections and above a certain depth in the Earth, have the same mass. It now seems certain that not all mountains are in isostatic equilibrum. This means that they are supported by the strength of the Earth's crust, not by floating—as are icebergs, which are in isostatic equilibrium.

Kirkwood gaps. Gaps in the distribution of asteriods corresponding to orbital periods that are commensurable with Jupiter's period.

Lipids. A class of substances, common in all living cells, that includes fats and related esters, On hydrolysis (see above), they yield such resulting substances as alcohols, sugars, and fatty acids.

Lithosphere. The solid, rocky, earthy part of the Earth in contrast to the barysphere (the heavy, deep interior) or the hydrosphere (the watery part at and near the surface).

Maar. A volcanic explosion crater that lacks a marked volcanic cone structure.

Mach's principle. That the inertia of a body is an aspect of its interaction with the matter in the rest of the universe.

Magma chamber. An underground chamber containing molten rock. Magma contains dissolved gasses (especially water vapor) which escape in volcanic eruptions. Thus, rock formed when lava cools differs in composition from the original magma.

Magnetopause. The boundary zone between the magnetosphere (see below) and space beyond, where the solar wind dominates.

Magnetosphere. That region of the upper atmosphere and near space dominated by the Earth's magnetic field. On the side facing the Sun, this region extends out to about a sixth of the distance to the Moon. On the dark side, the solar wind extends it into a long, cometlike tail.

Mare. The name (the Latin word for "sea") given by early astronomers to large, relatively smooth regions on the Moon that looked like seas as viewed through their telescopes. Tradition has led to the continued use of the term.

Maria. The plural of *Mare.*

Metamorphic. Refers to changes of a pronounced kind in rock caused largely by heat and pressure, but also by liquids and gases, usually requiring long periods of time.

Meteor. Originally this term referred to the phenomenon commonly called a "shooting star," in which a solid body from interplanetary space enters the Earth's atmosphere at high velocity and leaves an incandescent trail of vaporized matter. The term is now loosely extended to the bodies themselves as they orbit the Sun, although the word *meteoroids* is now becoming current. Bodies which reach the Earth are called meteorites; they are composed of metal, stone, or both, and range in size from subplanetary to microscopic.

Millibar. The unit of pressure in meteorology. It is one thousandth of a bar, which is one million dynes per square centimeter. Atmospheric pressure at sea level is one bar or about 14 pounds per square inch. In terms of millibars, standard atmospheric pressure at sea level is 1,013 millibars.

Mitochondria. Threadlike or granular particles (or organelles, as defined below), found in the protoplasm outside the nucleus of almost all cells, containing enzyme systems.

Mohorovičić discontinuity. Named after its discoverer, the Yugoslavian geophysicist A. Mohorovičić, who deduced a sudden increase in velocity of seismic waves at a depth of several tens of kilometers in the Earth. The Mohorovičić discontinuity, sometimes called Moho or M, is regarded as the boundary between the crust and the mantle of the Earth.

Moment of inertia. The product of the mass of a body and the square of the distance of the body from an axis of rotation. For bodies of appreciable size compared to the off-axis distance, the moment of inertia is the integral resulting from the sum of the moments of the infinitesimal bits of mass which the body comprises. The moment of inertia plays the same role for angular momentum that mass plays for linear momentum (see below).

Momentum. The product of the mass of a body and its linear velocity.

Mononucleotide. A nucleotide derived from three molecules (nitrogen compound, sugar, and phosphoric acid). Nucleotides are found in tissues and can be formed by partial hydrolysis of a nucleic acid.

NAD. The abbreviation for nicotinamide adenine dinucleotide, an enzymatic substance common in cells.

Nucleic acids. Long chain polymers of nucleotides. The latter consist of sugar phosphates tied to nitrogenous bases. There are two kinds of nucleic acids: (a) ribonucleic acid (RNA), which is metabolically highly active and seems to be linked to protein synthesis, and (b) desoxyribonucleic acid (DNA), which appears to be the carrier of genetic information.

Nucleotides. Compounds consisting of one or more units of phosphate-pentose-nitrogen base. They are found in tissues and result from the hydrolysis of nucleic acids.

Oligonucleotides. One form of nucleotides (see above).

Organelles. Subsystems of a cell which have distinct composition and function. They are specific particles of organized, living matter present in almost all cells. For example, mitochondria (see above) are common organelles of cells, characterized by their enzyme role.

Orogeny. The process of mountain-making, especially by foldings of the Earth's crust.

Peptides. Combinations of amino acids in which the amino group of one acid is tied to the carboxyl group of another. They may result from the hydrolysis of a protein. A dipeptide is a peptide that produces two molecules of amino acid on hydrolysis.

Peridotite. A coarse, granular, igneous rock.

Phytol. An oily alcoholic made by hydrolysis of chlorophyll.

Planck's radiation law. Expresses the fundamental notion that electromagnetic radiation has both wave and particle aspects and that it may be considered as made up of discrete packets or quanta of energy. Planck's law is basic to quantum mechanics and states that the energy of a quantum of radiation is proportional to frequency, or that energy equals frequency multiplied by a constant h, known as Planck's constant, the quantum of action.

Plasma. Ionized gas in which the number of positive and electric ions are about equal, so that the gas as a whole is essentially neutral.

Plutonic. Refers to igneous rocks formed at great depths within the Earth.

Polarization. When applied to electromagnetic waves, this term means that the vibration of electric or magnetic field vectors is confined to one plane. Polarization also refers to the separation of positive and negative electric charges by an electric field.

Polysaccharides. Carbohydrates that can be broken down by hydrolysis into simpler sugars.

Polynucleotides. See *Nucleotides.*

Polypeptides. Polyamides (crystalline compounds based on ammonia) that yield amino acids on hydrolysis.

Poynting-Robertson effect. The drag effect on a body revolving about the Sun, caused by solar radiation pressure. This minute effect acts to cause the body to spiral inwards, effective only for finely divided matter.

Pristane. A saturated liquid hydrocarbon obtained from the liver oils of some sharks.

Protonosphere. The name given to the region of the upper atmosphere above 1000 kilometers where the ions of hydrogen, or protons, are dominant particles (as against the ions of atomic oxygen at lower altitudes).

Purines. Crystalline carbon-hydrogen-nitrogen compounds made from uric acid.

Pyrimidines. Weak carbon-hydrogen-nitrogen bases that are a part of nucleotides.

Quantasome. Literally, a small and discrete body. The term is used in this book by Melvin Calvin in his chapter on chemical evolution of life to denote the ultimate biological unit of a cell that converts solar energy.

Rayleigh scattering. The scattering (or random reflection) of radiation by particles, such that the intensity of the scattered light is inversely proportional to the fourth power of the wavelength. The blue of the sky is due to the fact that air scatters the blue (short wavelength) component of sunlight much more effectively than the red (long wavelength) component.

Quasars ("quasi-stellar sources"). These are often intense radio emitters. Unlike most radio sources, which are associated with galaxies, quasars appear as starlike objects. They have extremely red-shifted spectra which, if interpreted in the same fashion as the red shift of galaxies, would make them the most distant objects observed. There is some question as to this interpretation because of the extreme brightness necessary to make them observable. Their origin and nature are not known.

Red Shift. See *Doppler effect.*

Relativistic. An adjective often used in the phrase "relativistic electron" or "relativistic particle" to indicate a velocity approaching that of light.

Relativity. Refers specifically to the theory advanced by Albert Einstein. In its first form, the Special Theory of Relativity, it was based on the hypothesis that the velocity of light is a universal constant, independent of the motion of the source or observer, and that all observers moving relative to each other in rectilinear nonaccelerated motion are equivalent. Several results ensue: that simultaneity is not an absolute concept, that energy has mass ($E=mc^2$), and that scales of space and time depend on relative velocity. The theory was later generalized to include accelerated motions (the General Theory), which led to the conclusion that the acceleration produced by gravitation is logically identical with other accelerations.

Reticulum. The network structure present in protoplasma.

Roche's limit. The distance from the center of a gravitating body (e.g., a planet) within which a fluid satellite would be pulled apart by the tide-raising forces of the gravitating body. It was first derived by the French mathematician E. A. Roche in 1850.

Shield. A continental crustal feature that has remained largely stable over a long period of time.

Stratigraphy. The study of the origin and chronology of rock layers.

Tectonic. Pertaining to the rock structure of the Earth and its crustal deformation.

Titius-Bode law. Discovered by J. D. Titius and published by J. E. Bode in 1772, the law is an empirical formula giving the mean distances of the planets from the Sun. While its values for seven of the planets and the asteroids are good approximations to observation, those for Neptune and especially Pluto are unsatisfactory.

Tropopause. The division between the stratosphere and the troposphere at a height ranging from 10 to 20 kilometers depending on latitude and season.

Tsunamis. Waves in the ocean, sometimes erroneously called tidal waves, usually caused by large earthquakes beneath the ocean bottom or coastal regions.

Whistlers. Electromagnetic waves, created by lightening strokes, which travel along magnetic lines of force from one hemisphere to another. In this process the frequencies in the original packet are drawn out so that the wave, converted to an audio signal, sounds like a whistle, starting high and ending low in pitch.

A Note on the Conversion of Metric-English Measure and Temperature Scales

The metric system is used in this book because of its international usage in science and because of its use in everyday affairs by most nations. The tables below provide units in both metric and English systems.

Angstrom (A)	$= 10^{-10}$ meter
Micron (μ)	$= 10^{-6}$ meter
Millimeter (mm)	$= 10^{-3}$ meter
Centimeter (cm)	$= 10^{-2}$ meter
Kilometer (km)	$= 10^{3}$ meters

The meter is an arbitrary unit, now defined in wavelengths. In the United States the relationship of inches to the meter is legally defined as 39.37 inches = 1 m. Some metric equivalents are as follows:

Millimeter (mm)	= 0.04 inch	Inch (in)	= 2.54 cm
Centimeter (cm)	= 0.39 inch	Foot (ft)	= 30.48 cm
Meter (m)	= 39.37 inches	Yard (yd)	= 91.44 cm
Kilometer (km)	= 0.62 miles	Mile (mi)	= 1.61 km

It is not difficult to sense the approximate English unit equivalents of linear metric measure if one remembers that a meter is about 3 feet or 1 yard (thus 2,000 meters equal about 2,000 yards or 6,000 feet, or a little more than a mile) and that a kilometer is about six-tenths of a mile (thus an altitude of 300 km is about 180 mi, 500 km about 300 mi, 1,000 km about 600 mi).

The two common metric mass units are the gram (g) and the kilogram (kg); the latter equals 1,000 g. A pound is equal to 453.6 grams, which is pretty close to half a kilogram. A kilogram is approximately 2.2 pounds. For ordinary reading, one can readily get the feel for the pound equivalent of kilograms by rounding up and multiplying by 2 (thus 460 kg can be rounded up to 500, which on multiplying by 2 gives 1,000 pounds).

The common temperature scales are the ordinary scales of Fahrenheit and Celsius (identical to the Centigrade scale) and the corresponding absolute Rankine and Kelvin scales. The last two start with zeros at absolute zero; the Fahrenheit scale starts with zero at 32° F below the freezing point of water, and Celsius with zero at that point. The table below brings out these relationships.

	Celsius or Centigrade	Kelvin	Fahrenheit	Rankine
Boiling Point of Water	100°C	373°K	212°F	672°R
Freezing Point of Water	0°C	273°K	32°F	492°R
Absolute Zero	−273°C	0°K	−460°F	0°R

This book uses the Celsius or the Kelvin scale, between which conversion is simple: to go from Celsius to Kelvin degrees, add 273; subtract 273 from °K to get °C. At very high temperatures, say the solar corona of one million degrees K, the difference between the two is not significant.

The relation between Celsius and Fahrenheit scales is straightforward: (a) the number of degrees between the freezing and boiling points of water on the Celsius scale is 100; on the Fahrenheit scale, 180; and so an interval of 1°F is equal to 5/9 of a degree C; (b) account must be taken of the 32° difference for the freezing point of water. To convert degrees C to Fahrenheit, one takes 9/5 of the degrees C and adds 32; to convert from Fahrenheit to Celsius, one first subtracts 32 and then takes 5/9 of the difference.

Index